List of Symbols

Symbol	Meaning	Symbol	Meaning
{ }	set	$m\angle A$	measure of angle A
$\in$	is an element of	$\triangle, \triangle s$	triangle(s)
$\notin$	is not an element of	$\text{int}\triangle$	interior of triangle
$=$	is equal to	$\text{ext}\triangle$	exterior of triangle
$\neq$	is not equal to	$\circ$	degree(s)
$\emptyset$	empty set	π	pi (approximately 3.1416)
$<$	is less than	rad	radian(s)
$>$	is greater than	$'$	minute(s)
$\leq$	is less than or equal to	$''$	second(s)
$\geq$	is greater than or equal to	$\perp$	is perpendicular to
$\subset$	is a subset of	⌐	perpendicular rays (in a diagram)
$\not\subset$	is not a subset of	$\longleftrightarrow$	corresponds to
:	such that	↮	does not correspond to
$\cup$	union	$\parallel$	is parallel to
$\cap$	intersection	$\square$	parallelogram
$|x|$	the absolute value of x	$\square$	rectangle
PQ	the distance from P to Q	$\sim$	is similar to
$A-B-C$	B is between A and C	$\sqrt{r}$	the square root of r
$\overline{AB}$	segment AB	$\odot$	circle
$\overleftrightarrow{AB}$	line AB	$\text{int}\odot$	interior of circle
$\overrightarrow{AB}$	ray AB	$\text{ext}\odot$	exterior of circle
$\cong$	is congruent to	$\overarc{AB}$	arc AB
$\ncong$	is not congruent to	$m\overarc{AB}$	measure of arc AB
$\angle, \angle s$	angle(s)	$\text{area}ABCDE$	area of polygon $ABCDE$
$\text{int}\angle$	interior of angle	$\text{area}\odot$	area of circle
$\text{ext}\angle$	exterior of angle		

GEOMETRY

for College Students

Second Edition

The Prindle, Weber & Schmidt Series in Mathematics

Swokowski, *Calculus with Analytic Geometry, Fifth Edition*
Swokowski, *Calculus with Analytic Geometry, Fourth Edition (Late Trigonometry)*
Swokowski, *Calculus of a Single Variable*
Swokowski, *Fundamentals of College Algebra and Trigonometry, Seventh Edition*
Swokowski, *Fundamentals of College Algebra, Seventh Edition*
Swokowski, *Fundamentals of Trigonometry, Seventh Edition*
Swokowski, *Precalculus: Functions and Graphs, Sixth Edition*
Tan, *Applied Calculus, Second Edition*
Tan, *Applied Finite Mathematics, Third Edition*
Tan, *Calculus for the Managerial, Life, and Social Sciences, Second Edition*
Tan, *College Mathematics, Second Edition*
Trim, *Applied Partial Differential Equations*
Venit and Bishop, *Elementary Linear Algebra, Third Edition*
Venit and Bishop, *Elementary Linear Algebra, Alternate Second Edition*
Wiggins, *Problem Solver for Finite Mathematics and Calculus*
Willard, *Calculus and Its Applications, Second Edition*
Wood and Capell, *Arithmetic*
Wood, Capell, and Hall, *Developmental Mathematics, Fourth Edition*
Wood and Capell, *Intermediate Algebra*
Zill, *A First Course in Differential Equations with Applications, Fourth Edition*
Zill, *Calculus with Analytic Geometry, Second Edition*
Zill, *Differential Equations with Boundary-Value Problems, Second Edition*

The Prindle, Weber & Schmidt Series in Advanced Mathematics

Brabenec, *Introduction to Real Analysis*
Ehrlich, *Fundamental Concepts of Abstract Algebra*
Eves, *Foundations and Fundamental Concepts of Mathematics, Third Edition*
Keisler, *Elementary Calculus: An Infinitesimal Approach, Second Edition*
Kirkwood, *An Introduction to Real Analysis*
Ruckle, *Modern Analysis: Measure Theory and Functional Analysis with Applications*

To Sharon

GEOMETRY
for College Students

Second Edition

Peter B. Geltner
Santa Monica College

Darrell J. Peterson
Santa Monica College

PWS-KENT PUBLISHING COMPANY, BOSTON

PWS-KENT
Publishing Company

20 Park Plaza
Boston, Massachusetts 02116

PWS-KENT Publishing Company is a division of Wadsworth, Inc.

Library of Congress Cataloging-in-Publication Data

Geltner, Peter B.
    Geometry for college students / Peter B. Geltner, Darrell J.
Peterson. — 2nd ed.
      p. cm.
    Includes index.
    ISBN 0-534-92472-7
    1. Geometry, Plane. I. Peterson, Darrell J.   II. Title.
QA455.G45 1990
516.22—dc20   90-48023
                    CIP

Sponsoring Editor: Tim Anderson
Assistant Editor: Diana Kelley
Production Editor: Laura Mosberg
Manufacturing Coordinator: Margaret Sullivan Higgins
Interior Design: IPS Publishing, Inc., Laura Welch / Laura Mosberg
Cover Design: Laura Mosberg
Cover Photo: Slide Graphics
Typesetting: Publication Services, Inc.
Cover Printing: John P. Pow, Company, Inc.
Printing and Binding: The Maple-Vail Book Manufacturing Group

Printed in the United States of America.
91 92 93 94 95 – 10 9 8 7 6 5 4 3 2 1

CONTENTS

PREFACE

This second edition of *Geometry for College Students* features the precise coverage of the elementary concepts of plane Euclidean geometry, which made the first edition so successful with faculty and thousands of student users. It is written specifically with the college student in mind, not as an attempt to rework a high school oriented text. The book is designed for individuals lacking skills in geometry, especially those college students who have had little or no formal background in the subject. The book can also serve as a brief overview of geometry and has proven useful for the mathematical education of prospective teachers.

Chapters 1 through 7 present the essential material of plane geometry, and can easily be covered in a three-unit, one-semester course, perhaps omitting the optional trigonometry section. The additional topics in Chapters 8, 9, and 10, provide enrichment materials and enable the book to be used for a five-unit, one semester course, or for a two-quarter course. These three chapters are sufficiently independent so that any of them could be used separately.

Experience in using high school texts for college students indicates that many students are overwhelmed with a large quantity of data covered in too short a time. High schools commonly spend a full year on geometry, whereas most colleges offer only one semester. The material in this text is treated in a precise manner that helps students learn, understand, and apply the correct terminology in the shortest possible time. Versions of this text have been class-tested extensively, and the withdrawal rate of students in these classes has significantly declined.

A modern approach is used for definitions and notation. The deductive method of proof is carefully developed in order to ease the student into the new concept. Sets are used, only as necessary, to help produce clear definitions.

The exercise sets are long and varied to give the user a wide choice of assignments. Many applications and computational problems are included, as well as theorems and other statements to be proved. Constructions are spaced throughout the book to demonstrate that they are an integral part of geometry, rather than an isolated topic. Each chapter concludes with a chapter review, a concept list (including page references), and an exercise set.

New in the Second Edition

In response to suggestions from several users and reviewers of the first edition we have incorporated the following changes:

- A new, two-color interior design offers a cleaner and more readable appearance.

- Many of the definitions, including the definition of region, have been rewritten for even greater clarity.

- New examples have been added throughout the text.

- The addition of 150 new exercises, including many application problems and constructions, provides further illustrations of the concepts.

- Several proofs of the Pythagorean Theorem have been included in Exercise Set 7.2.

- A discussion of coordinate geometry in proofs has been added to Section 8.3.

- An entirely new section (8.5) on transformations, rotations, and reflections has been added.

Supplements

This second edition is complemented by an up-to-date set of resource materials, including:

Instructor's Manual with answers to all even numbered problems.

Test Bank with three sample tests for each chapter.

Acknowledgments

We wish to express our appreciation to the students of Santa Monica College who helped class-test many portions of the book for several years, and to the faculty members of the Santa Monica College Mathematics Department, who encouraged us and gave us helpful suggestions.

We also wish to thank the following reviewers who helped us make changes in this second edition that increased quality and usefulness of the text: Kathleen Bavelas, *Manchester Community College*; Elise Grabner, *Slippery Rock University*; Virginia Hanks, *Western Kentucky University*; Karen Hinz, *Anoka-Ramsey Community College*; Paul A. Kennedy, *South-*

west *Texas State University*; Carl M. Kerns, *Mesa State College*; Vivien Miller, *Mississippi State University*; Rose Strahan, *Delta State University*; and Lenard E. Taylor, *Sierra Community College*.

We are grateful for the valuable assistance of the staff at PWS-KENT Publishing Company, including Tim Anderson, Cathie Griffin, Barbara Lovenvirth, Diana Kelley, and Laura Mosberg.

Peter B. Geltner
Darrell J. Peterson

Introduction

Courtesy of Culver Pictures

Euclid (ca. 300 B.C.)

Historical Note

In ancient times, mathematicians tried to define the terms *point* and *line*. Pythagoras defined a point as "a monad having position," Plato as "the beginning of a line," and Euclid as "that which has no part." Some early definitions of line were "that of which the middle covers the end," or "length without breadth," or "the flux of a point." We now consider these and some other basic terms as undefined, and we build definitions of other terms upon them.

In this chapter, we introduce the basic elements of a logical system, which include undefined terms, definitions, and postulates. We also introduce some simple geometric objects and discuss the relationships between these objects and real numbers through the concept of length and measure. Constructions are included at the end of the chapter. We defer the discussion of theorems until Chapter 2.

1.1 A LOGICAL SYSTEM

Geometry is the study of the properties and characteristics of certain sets, such as lines, angles, triangles, and circles. The subject is developed carefully and logically by what is known as *deductive reasoning*. A system that depends on deductive reasoning is known as a *logical system*.

A **logical system** consists of undefined terms, definitions, assumptions, and theorems. The undefined terms we use are **set, point, line,** and **plane**. We do not define these terms because, in the attempt to define them, we are forced to use other terms that have not been defined. For example, a *set* may be thought of as a group or collection of objects; however, the terms *group* and *collection* have not been defined, so we may not use these words to define *set*. A *point* may be thought of as a location in space or an object with no dimension (length, width, or height); however, we may not use the words *location, space,* and *dimension*. Likewise, a *line* may be thought of as a one-dimensional straight object extending infinitely far in two directions and a *plane* may be thought of as a two-dimensional flat object with no boundaries. (Can you list all the undefined words used in the last sentence?)

In Figure 1.1, A represents a point, j represents a line, and P represents a plane.

Figure 1.1

We shall use braces, $\{\ \}$, to indicate sets. For example, $\{1, V, \$\}$ is the set consisting of the numeral one, the letter V, and a dollar sign. The objects within a set are **elements** of the set. Often, sets are given names. If we call the above-mentioned set S, we write $1 \in S$, meaning "1 is an element of S." The symbol $\notin$ means "is not an element of." Thus, $2 \notin S$. In general, a slash through a symbol means *not*. Thus, $\neq$ means "is not equal to."

■

Example 1

For $A = \{3, y, /\}$, insert the proper symbol ($\in$ or $\notin$):

(a) $y ? A$ (b) $[? A$ (c) $A ? A$

Solution

(a) $y \in A$, since y is an element of A.
(b) [$\notin A$, since [is not an element of A.
(c) $A \notin A$, since A is not an element of A.

∎

A set is **finite** if the number of elements in the set is equal to some nonnegative integer n. Otherwise, the set is **infinite**. Thus, $\{1, 2, 3\}$ is a finite set, whereas $\{$integers$\}$ is infinite, because $\{$integers$\}$ means *all* integers unless otherwise noted.

∎

Example 2

Are the following sets finite or infinite?

(a) $H = \{$hairs on a monkey$\}$
(b) $\{$all possible fractions$\}$

Solution

(a) H is finite, since the number of hairs on a monkey equals some nonnegative integer, even though that integer may be a very large number.
(b) The set of *all possible fractions* is infinite, since no nonnegative integer can be assigned to the number of fractions. No matter what integer we attempt to pick, there would be a greater number of fractions than that integer.

∎

Two sets A and B are **equal** if they contain the same elements. For example, if $A = \{1, 2\}$ and $B = \{1, 2\}$, A and B are equal; we write $A = B$. Two sets A and B are in **one-to-one correspondence** if for each element of A there is exactly one element of B, and for each element of B there is exactly one element of A. Thus, the sets $R = \{3, 4\}$ and $S = \{*, \&\}$ are in one-to-one correspondence, but $R \neq S$.

∎

Example 3

Which of the following sets are equal and which are in one-to-one correspondence?

$A = \{5, 7, 9\}$ $B = \{$L, O, G$\}$ $C = \{$G, O, L$\}$ $D = \{7, 5\}$

Solution

Sets B and C are equal since they contain exactly the same elements. Order of the elements is not important. Sets A and B, sets A and C, and sets B and C are in one-to-one correspondence.

∎

A set is **empty** if it has no elements. Thus, the set of four-headed dogs and the set of five-nosed people are both empty. Furthermore, these two sets are equal, since they contain the same elements, namely no elements. Thus, it follows that any set that is empty must equal any other set that is empty. There is only one empty set, although there are many ways to describe it. The empty set is denoted by $\emptyset$.

The set A is a **subset** of the set B if every element in A is also an element of B. We write $A \subset B$. Thus $\{1\} \subset \{1, 2\}, \{2, 3\} \subset \{2, 3\}, \{3\} \subset \{3\}$, and $\emptyset \subset \{4, 7\}$, but $\{2\} \not\subset \{1, 3\}, \{2\} \not\subset \emptyset$, and $\{1, 2, 3\} \not\subset \{1, 2, 4, 7\}$. In general, every set is a subset of itself and $\emptyset$ is a subset of every set.

Example 4

For $A = \{3, y, /\}$, insert the proper symbol ($\subset$ or $\not\subset$), given that $B = \{3, y\}$ and $C = \{y, *\}$.

(a) $A\ ?\ B$ (b) $B\ ?\ A$ (c) $C\ ?\ A$ (d) $\emptyset\ ?\ A$

Solution

(a) $A \not\subset B$ since $/$ is in A but not in B.
(b) $B \subset A$ since every element in B is also in A.
(c) $C \not\subset A$ since C contains $*$, and A does not.
(d) $\emptyset \subset A$ since $\emptyset$ is a subset of every set.

∎

Sometimes a set is defined using "set builder" notation. For example, $\{x: x < 3\}$ is read "the set of all x such that x is less than 3." Using this notation, we define the **intersection** ($\cap$) of A and B as $A \cap B = \{P: P \in A$ and $P \in B\}$ and the **union** ($\cup$) of A and B as $A \cup B = \{P: P \in A$ or $P \in B\}$.

The word *or* means that the element is in A or in B or in both. The word *and* means that the element is in both. If $A = \{1, 2, 3, 4\}$ and $B = \{4, 6, 7\}$, then $A \cup B = \{1, 2, 3, 4, 6, 7\}$ and $A \cap B = \{4\}$.

Example 5

If $C = \{3, 5, 6\}$ and $D = \{*, ?, 6\}$, find

(a) $C \cup D$ (b) $C \cap D$ (c) $\emptyset \cup C$ (d) $\emptyset \cap C$

Solution

(a) $C \cup D = \{3, 5, 6, *, ?\}$
(b) $C \cap D = \{6\}$
(c) $\emptyset \cup C = C$
(d) $\emptyset \cap C = \emptyset$

Many of the examples used in this text are concerned with applications to the world around us. These applications are sometimes called *word problems* or *stated problems* in various algebra textbooks. Although it is assumed that the reader has had experience with application problems in a basic algebra course, we include some review guidelines here. Some of the steps may not be necessary in a particular problem you are working.

Guidelines for solving application problems

1. Read the problem carefully, several times if necessary, to be certain that you understand the meaning of all the words and to comprehend the situation being described.

2. Determine which facts are given.

3. Determine which information is to be found.

4. Assign variable names to any unknown quantities that are needed in the problem.

5. Sketch a diagram, when possible, to assist you in analyzing the problem.

6. Form an equation containing the variables, to translate the problem from English to algebra.

7. Solve the equation and use the solution to determine all the requested information.

8. Check all answers by determining if they correctly answer all the questions asked in the *original problem*.

■

Example 6

A president of a college wishes to determine the number of different students in a special scholars program consisting of one English class and one mathematics class. The number of students in the English class is 36 and the number of students in the mathematics class is 42. The number of students that are taking both classes is 25. How many different students are in the special program?

Solution

Figure 1.2

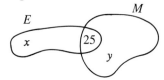

E

x 25 M

y

Let E be the set of students in the English class and M be the set of students in the mathematics class. Then $E \cap M$ will be the set of students in both classes. Thus, the set $E \cap M$ contains 25 students. Let x = number of students in English but not in math. Let y = number of students in math but not in English. See Figure 1.2. Since $E \cap M$ is a subset of both E and M, it follows that $x + 25 = 36$ and $y + 25 = 42$. Solving for x and for y, we obtain $x = 11$ and $y = 17$. Thus, $11 + 25 + 17 = 53$ students are in the special program. Notice that the number of students in $E \cup M$ equals the number of students in E, plus the number of students in M, minus the number of students in $E \cap M$; namely $36 + 42 - 25 = 53$.

■

EXERCISE SET 1.1

In Exercises 1–6, state whether each of the sets is finite or infinite.

1. The even integers

2. The presidents of the United States

3. All possible steamship routes from San Francisco to Tokyo

4. The grains of sand on the beach at Waikiki

5. The babies born in City Hospital in 1985

6. The three-headed kangaroos that wear bow ties

7. Give three descriptions of the empty set.

8. Which of the sets $\{1, 4, 9\}, \{a, b, c\}, \{1^2, 2^2, 3^2\}, \{1, 2, 3, 4\}, \{a^2, b^2, c^2\}$, $\emptyset, \{\sqrt{1}, \sqrt{16}, \sqrt{81}\}$, and $\{9, 1, 4\}$ are equal? Which are in one-to-one correspondence?

In Exercises 9–29, let $A = \{1, 2, 3, 4\}$, $B = \{x: x$ is a positive integer $\}$, $C = \{2, 4\}$, $D = \{1\}$, $E = \{\triangle, \square\}$, and $F = \{2, 4, \triangle, \square\}$.

In Exercises 9–15, insert the proper symbol ($\subset$ or $\not\subset$).

9. C ? A

10. D ? A

11. A ? B

12. D ? C

13. $\emptyset$? E

14. F ? B

15. E ? F

In Exercises 16–22, find:

16. $C \cup D$

17. $A \cup B$

18. $A \cap C$

19. $A \cap D$

20. $C \cap E$

21. $C \cup E$

22. $A \cap (C \cup D)$

In Exercises 23–29, insert the proper symbol ($\in$ or $\notin$).

23. 4 ? A

24. 3 ? B

25. 3 ? F

26. $\triangle$? B

27. $\triangle$? $(D \cup E)$

28. 3 ? $(A \cap C)$

29. $\sqrt{4}$? C

In Exercises 30–37, let $A = \{x : x$ is a positive integer$\}$, $B = \{y : y$ is an even integer$\}$, $C = \{z : z = 3n, n$ is a positive integer$\}$, and $D = \{99, 108, 111, 127\}$.

In Exercises 30–33, insert the proper symbol ($\subset$ or $\not\subset$).

30. A ? B

31. C ? A

32. B ? C

33. D ? C

In Exercises 34–37, find:

34. $B \cup D$

35. $A \cup C$

36. $A \cap B$

37. $C \cap D$

In Exercises 38–41, insert the proper symbol ($\in$ or $\notin$).

38. If $E = $ {all breeds of dogs}, then a collie ? E.

39. If $F = $ {all breeds of cats}, then a Pekinese ? F.

40. If $G = $ {the 1990 football team of Superville College}, then a 1990 shortstop for Superville College ? G.

41. If $H = $ {all winners of Oscar awards}, then Meryl Streep ? H.

42. A professor in a community college wishes to determine the number of different students in a program consisting of one history class and one typing class. The number of students in the history class is 27, and the number of students in the typing class is 15. The number of students that are taking both classes is 8. How many different students are in the program?

43. A doctor in a hospital wishes to determine the number of different patients in a hospital ward. In the ward, there are 75 patients who have measles, 56 patients who have chicken pox, and 10 patients who have both chicken pox and measles. No other patients are in the ward. How many different patients are in the ward?

44. A geometry class contains 8 left-handed women and 29 right-handed women. The number of left-handed students in the class is 13 and the number of right-handed students in the class is 35. How many students are in the geometry class?

45. An algebra class contains 18 left-handed men and 12 right-handed men. The number of left-handed students in the class is 23 and the number of right-handed students in the class is 15. How many students are in the algebra class?

46. A shelf contains 22 red books, 15 history books, 11 large books, 9 red history books, 6 large red books, 7 large history books, and 2 large red history books. How many different books are on the shelf?

47. A pet store contains 12 brown dogs, 10 spaniels, 11 large dogs, 7 brown spaniels, 4 large brown dogs, 5 large spaniels, and 3 large brown spaniels. How many different dogs are in the pet store?

48. In a certain parking lot there are 30 two-door cars, 17 foreign cars, and 25 white cars. There are 17 white two-door cars, 13 foreign two-door cars, 12 white foreign cars, and 8 white foreign two-door cars. What is the total number of cars on the lot?

49. In a forest there are 385 tall trees with red leaves, 407 hardwood trees with red leaves, 462 tall hardwood trees, and 300 tall hardwood trees with red leaves. There are 758 tall trees, 537 trees with red leaves, and 737 hardwood trees. How many trees are in the forest?

1.2 POSTULATES

Space is defined as the set of all points. By this definition, if an object is a point, then it is in space. However, nothing that has been done so far has suggested how many points exist. Furthermore, no relationship has been stated between the undefined words *point*, *line* and *plane*. In order to use these words, we must make some assumptions about the relationships among them.

In algebra, assumptions are called **axioms**. These are reviewed in Appendix A. In geometry, assumptions are called **postulates**.

Postulate 1.1	Every line contains at least two distinct points.

Postulate 1.2	Any two distinct points in space have exactly one line that contains them.

Postulate 1.3	Every plane contains at least three distinct points, not all on one line.

Postulate 1.4	Any three distinct points in space not on one line have exactly one plane that contains them.

Postulate 1.5	For any two distinct points in a plane, the line containing the points is also in the plane.

Postulate 1.6	No plane contains all points of space.

Postulate 1.7	(**Ruler Postulate**) There is a one-to-one correspondence between the set of all real numbers and the set of all points on a line.

The number that corresponds to each given point on a line is the **coordinate** of the point. Since the set of real numbers is infinite, Postulate 1.7 implies that a line contains infinitely many points. Therefore, we no longer need Postulate 1.1; however, at the time Postulate 1.1 was introduced, it was needed to describe the assumed differences between a point, a line and a plane. In fact, we can see by Postulate 1.5 that a plane must also contain infinitely many points. Statements that can be derived from postulates and definitions are **theorems**. Theorems and their proofs are discussed in the next chapter.

Example 1

Which postulate applies to the following situation?

Given three distinct points A, B, and C on a coordinate line, if A has coordinate -3 and B has coordinate 5, then C cannot have 5 as a coordinate. Also, C cannot have -3 as a coordinate.

Solution

Postulate 1.7, the Ruler Postulate. Since there is a one-to-one correspondence between the real numbers and points on a line, the coordinate 5 cannot be assigned to the two distinct points B and C, and the coordinate -3 cannot be assigned to the two distinct points A and C.

■

Figure 1.3

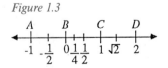

In using the Ruler Postulate, one commonly picks a correspondence in which the coordinates are equally spaced along a number line, as in Figure 1.3.

The **absolute value** of a number x, denoted by $|x|$, is defined by $|x| = x$ if $x \geq 0$ and $|x| = -x$ if $x < 0$.

Example 2

Find $|\,3\,|, |\,0\,|, |\,-2\,|, |\,c + 1\,|$, and $|\,y - 5\,|$.

Solution

$|\,3\,| = 3$, since $3 \geq 0$,
$|\,0\,| = 0$, since $0 \geq 0$,
$|\,-2\,| = -(-2) = 2$, since $-2 < 0$,
$|\,c + 1\,| = c + 1$ if $c + 1 \geq 0$ and $|\,c + 1\,| = -(c + 1)$ if $c + 1 < 0$,
$|\,y - 5\,| = y - 5$ if $y \geq 5$ and $|\,y - 5\,| = -(y - 5) = 5 - y$ if $y < 5$.

In the solution above we used the statement $y \geq 5$ instead of $y - 5 \geq 0$ and the statement $y < 5$ instead of $y - 5 < 0$. The addition axiom of inequality allows us to add 5 to both sides of the inequalities, thereby simplifying the expressions.

If x and y are coordinates of points P and Q, respectively, then the **distance** from P to Q is $|x - y|$. We denote the distance from P to Q by PQ. That is, $PQ = |x - y|$.

Distance is an important concept in geometry since it is used in many definitions. For example, a **circle** is the set of all points in a plane at a given distance from a given point. The given point is the **center** of the circle, and the given distance is the **radius** of the circle. (The plural of radius is **radii**.) If the given point is P, then the circle is called *circle P.*

Definition | Point B is **between** points A and C if A, B, and C are on the same line and $AB + BC = AC$. We write $A-B-C$.

Example 3

In Figure 1.4, indicate whether the statements $A-B-C$, $A-C-B$, and $A-D-C$ are true or false.

Solution

Figure 1.4

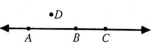

$A-B-C$ is true, since A, B, and C are on the same line and $AB + BC = AC$. $A-C-B$ and $A-D-C$ are both false, since $AC + CB \neq AB$ and A, D, and C are not all on the same line.

EXERCISE SET 1.2

In Exercises 1–5, which postulate is illustrated by each of the following, if A, B, and C are distinct points, k and m are lines, and P is a plane?

1. If $A \in P$, $C \in P$, $A \in k$, and $C \in k$, then $k \in P$.

2. If $B \in k$, $C \in k$, and $k \neq m$, then B and C cannot both be on m.

3. If $A \in k$, $B \in k$, $A \in m$, and $B \in m$, then $k = m$.

4. There is a point C such that $C \notin P$.

5. If $A \in k, B \in k, C \in m$, and $C \notin k$, then there is a plane P such that $A \in P, B \in P$, and $C \in P$.

6. Draw a number line, as in Figure 1.3, and mark the points on the line that represent each of the following: 0, 3, −2, 3/2, −1/3, √3, and π.

In Exercises 7–22, find the values:

7. $\lvert -7 \rvert$	13. $\lvert d \rvert$	18. $\lvert x - 3 \rvert$ if $x < 3$
8. $\lvert -23 \rvert$	14. $\lvert -t \rvert$	19. $\lvert a - b \rvert$ if $a > b$
9. $\lvert -0 \rvert$	15. $\lvert x^2 \rvert$	20. $\lvert a - b \rvert$ if $a < b$
10. $\lvert 0 \rvert$	16. $\lvert -y^2 \rvert$	21. $\lvert x + y \rvert$ if $x \leq -y$
11. $\lvert 52 \rvert$	17. $\lvert x - 3 \rvert$ if $x > 3$	22. $\lvert x + y \rvert$ if $x > -y$
12. $\lvert 24 \rvert$		

In Exercises 23–30, find each of the indicated distances for the following diagram.

23. EF	25. BE	27. CA	29. CE
24. AB	26. EB	28. AC	30. DF

Figure for Exercises 23–30

31. If P and Q are distinct points on a line k with coordinates p and q, respectively, why does $PQ = QP$?

32. On a map, a number line is placed in such a way that a golf course with coordinate −10, a lake with coordinate 53, and a museum with coordinate 24 all fall on the line. All units on the map are in miles. Find the distance between the golf course and the lake and the distance between the lake and the museum.

33. A stellar map shows that three stars, Alpha, Beta, and Gamma, all fall on a straight line. Place a number line over the stars on the map. Alpha has coordinate -5, Beta has coordinate 6, and Gamma has coordinate 22. All units on the map are in parsecs. Find the distance between Alpha and Beta and the distance between Beta and Gamma.

34. What postulate shows that a three-legged chair is stable?

35. Why is a four-legged table not always stable?

1.3 SEGMENTS, RAYS, ANGLES, TRIANGLES

The following concept refers to the set of points represented by the union of points A and B and the set of all points between A and B. In general, definitions involving sets will be given in symbols.

Definition | The **line segment** $AB = \{A, B\} \cup \{P: A\text{–}P\text{–}B\}$. We write $\overline{AB}$.

Figure 1.5

Figure 1.5 is a diagram of $\overline{AB}$. Points A and B are the **endpoints** of $\overline{AB}$. From the definition we can see that $\overline{AB}$ and $\overline{BA}$ represent the same line segment. Notice that in Figure 1.1 a line is represented with arrows on the ends; the line containing A and B is denoted by $\overleftrightarrow{AB}$, in contrast to $\overline{AB}$ (line segments have no arrows).

The **length** of $\overline{AB}$ is the distance between its endpoints, denoted by AB.

Definition | Two line segments $\overline{AB}$ and $\overline{CD}$ are **congruent** if $AB = CD$. We write $\overline{AB} \simeq \overline{CD}$.

Figure 1.6

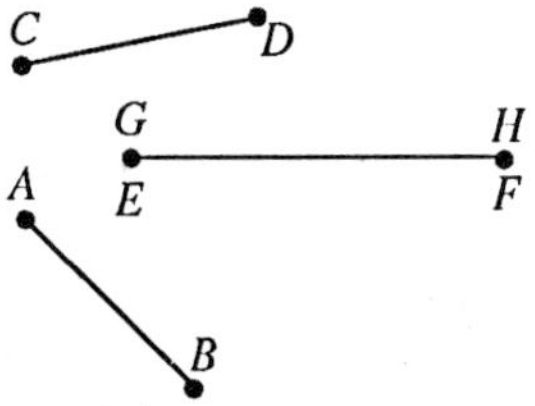

Example 1

In Figure 1.6, given that $\overline{AB}$ and $\overline{CD}$ have the same length, determine whether the following are true or false.

(a) $\overline{AB} \simeq \overline{CD}$ (b) $AB = CD$ (c) $\overline{AB} = \overline{CD}$ (d) $\overline{EF} \simeq \overline{GH}$
(e) $EF = GH$ (f) $\overline{EF} = \overline{GH}$

Solution

(a) true   (b) true   (c) false   (d) true   (e) true   (f) true

∎

$\overline{AB} = \overline{CD}$ has a different meaning than $\overline{AB} \simeq \overline{CD}$. The former means that the point sets are identical (equal); the latter means that the point sets are congruent. Thus, in Figure 1.6, $\overline{AB} \simeq \overline{CD}$, but $\overline{AB} \neq \overline{CD}$. Notice that point E is also named point G and point F is also named point H. Although not usually done, two names are used here to demonstrate the difference between equality and congruence of line segments.

Definition

> The **ray** $AB = \overline{AB} \cup \{P: A{-}B{-}P\}$. We write $\overrightarrow{AB}$.

Figure 1.7

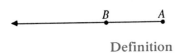

Point A is the endpoint of $\overrightarrow{AB}$. See Figure 1.7. Notice the use of the arrow.

Definition

> The **angle** $BAC = \overrightarrow{AB} \cup \overrightarrow{AC}$. We write $\angle BAC$.

Rays $\overrightarrow{AB}$ and $\overrightarrow{AC}$ are **sides** of $\angle BAC$. Point A is the **vertex** of $\angle BAC$. (The plural of vertex is vertices.)

Definition

> If $B{-}A{-}C$, then $\angle BAC$ is a **straight angle**.

In Figure 1.4, $\angle ABC$ is a straight angle with the vertex at B.

Figure 1.8

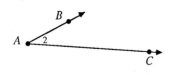

In Figure 1.8, the angle represented can be given different names, all representing the same set. From the definition we can see that $\angle BAC$ and $\angle CAB$ represent the same angle. When there is no possibility of confusion in a diagram, we can call the angle $\angle A$. Sometimes in a complicated diagram, it is easier to name an angle by using a number. The number is placed as shown in Figure 1.8, namely $\angle 2$.

Definition

> Given points R and S and line k, R and S are **on the same side of** k if $\overline{RS} \cap k = \emptyset$, and R and S are **on opposite sides of** k if $\{R\} \cap k = \emptyset$, $\{S\} \cap k = \emptyset$, and $\overline{RS} \cap k \neq \emptyset$.

■
Example 2

Figure 1.9

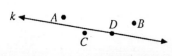

In Figure 1.9, A and B are on the same side of k, but A and C are on opposite sides of k. Why? Notice that A and D are not on the same side of k and also not on opposite sides of k. Why?

Solution

Points A and B are on the same side of k, since $\overline{AB} \cap k = \emptyset$. Points A and C are on opposite sides of k, since A and C are not in k and $\overline{AC} \cap k \neq \emptyset$. Points A and D are not on the same side of k and not on opposite sides of k, since D is in k.

∎

Definition

The **interior** of $\angle BAC = \{P: \angle BAC$ is not a straight angle, points P and C are on the same side of $\overleftrightarrow{AB}$, and points P and B are on the same side of $\overleftrightarrow{AC}\}$. We write int$\angle BAC$.

Definition

The **exterior** of $\angle BAC = \{P: P$ is not in $\angle BAC$ and not in int$\angle BAC\}$. We write ext$\angle BAC$.

We may consider the set of all points on only one side of the line containing a straight angle as the interior of the angle. The set of all points on the opposite side would then be the exterior of the angle.

∎

Example 3

Figure 1.10

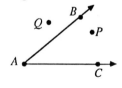

In Figure 1.10, determine whether P and Q are in the interior or exterior of $\angle BAC$.

Solution

$P \in$ int$\angle BAC$, since points P and C are on the same side of $\overleftrightarrow{AB}$ and points P and B are on the same side of $\overleftrightarrow{AC}$.

$Q \in$ ext$\angle BAC$, since Q is not in $\angle BAC$ and not in int$\angle BAC$.

∎

Definition

Triangle $ABC = \overline{AB} \cup \overline{BC} \cup \overline{AC}$, where A, B, and C are points not all on the same line. We write $\triangle ABC$. Segments $\overline{AB}$, $\overline{BC}$, and $\overline{AC}$ are **sides** of $\triangle ABC$.

Figure 1.11

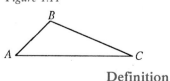

See Figure 1.11 for a diagram of $\triangle ABC$.

Definition

The **interior** of $\triangle ABC = \{P: P \in$ int$\angle ABC \cap$ int$\angle BAC\}$. We write int$\triangle ABC$.

Definition | The **exterior** of $\triangle ABC = \{P: P$ is not in $\triangle ABC \cup \text{int}\triangle ABC\}$. We write ext$\triangle ABC$.

In Figure 1.12, $P \in \text{int}\triangle ABC$, $Q \in \text{ext}\triangle ABC$, $A \in \triangle ABC$, $A \notin \text{int}\triangle ABC$, and $A \notin \text{ext}\triangle ABC$. Thus, P is not in $\triangle ABC$ but is in int$\triangle ABC$. In $\triangle ABC$, side $\overline{AB}$ and $\angle C$ are **opposite** each other.

Figure 1.13

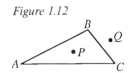

Figure 1.12

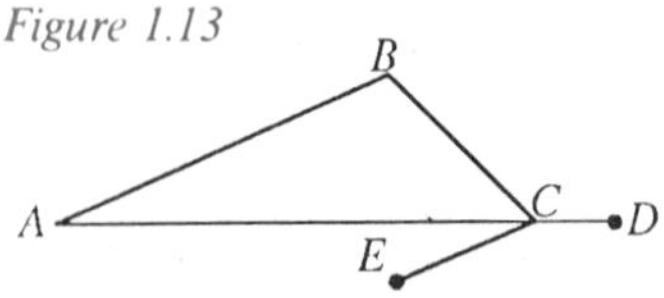

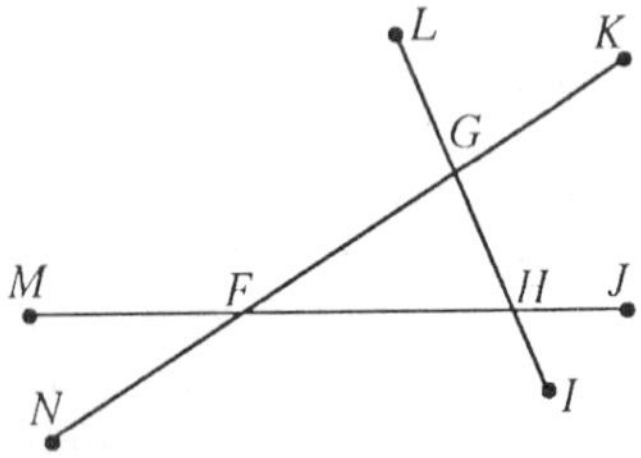

If we are given $\triangle ABC$ and a point D such that $A-C-D$, then $\angle BCD$ is an **exterior angle** of $\triangle ABC$. In Figure 1.13, angles A, B, and ACB are interior angles of the triangle. Given exterior $\angle BCD$, $\angle A$ and $\angle B$ are remote interior angles. Notice that $\angle DCE$ is not an exterior angle, since it is not true that $B-C-E$. Angles FHI, GHJ, HGK, FGL, GFM, and HFN are exterior angles of $\triangle FGH$.

EXERCISE SET 1.3

In Exercises 1–3, given the diagram below with $E-A-C$ and $F-A-B$, list all the:

1. Line segments

2. Lines

Figure for Exercises 1–3

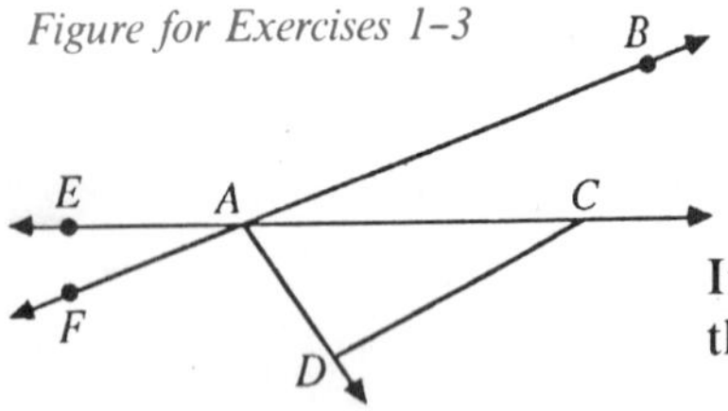

3. Rays

In Exercises 4–9, given the following diagram with $AB = EF$, insert the correct symbol $(=, \simeq, \neq, \not\simeq)$.

4. AB ? CD

5. $\overline{AB}$? $\overline{CD}$ (two answers)

6. CD ? EF

7. $\overline{CD}$? $\overline{EF}$ (two answers)

8. CD ? EG

9. $\overline{AB}$? $\overline{EG}$ (two answers)

Figure for Exercises 4–9

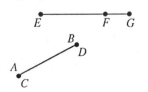

Figure for Exercises 10-19

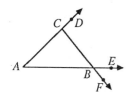

In Exercises 10–19, given the diagram with A–C–H:

10. List all the angles using three letters for each.

11. List all the points that are vertices of angles.

In Exercises 12–16, which of the following pairs of points are on the same side of $\overleftrightarrow{BC}$? Which are on opposite sides of $\overleftrightarrow{BC}$?

12. D and F 14. F and G 16. A and H

13. D and H 15. A and F

In Exercises 17–19, which labeled point or points are in the interior of each?

17. $\angle BCH$ 18. $\angle ACB$ 19. $\angle FCH$

Figure for Exercises 20–22

In Exercises 20–22, use the diagram below with A–B–E, C–B–F, and A–C–D:

20. Name the interior angles.

21. Name the exterior angles.

22. Name the remote interior angles for each exterior angle listed in Exercise 21.

23. Define *ray* in words.

In each of the Exercises 24-27, list five objects that remind you of the indicated geometric sets.

24. Line segments

25. Rays

26. Angles

27. Triangles

1.4 ANGLE MEASURE

Given a line in a plane, three distinct point sets are formed, namely the set of points on one side of the line, the set of points on the other side of the line, and the set of points forming the line itself. A **half-plane** is the set of all points on one side of a line. The line is the **edge** of the half-plane. A **closed half-plane** is the union of a half-plane and its edge.

Postulate
1.8

(Degree Protractor Postulate) Let $\overrightarrow{AB}$ be a ray on the edge of a closed half-plane H. For every ray $\overrightarrow{AP}$ such that $P \in H$ and $P \notin \overrightarrow{AB}$, there is exactly one real number r such that $0 < r \le 180$.

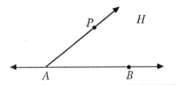

The real number in Postulate 1.8 is the **measure** of $\angle PAB$, denoted by m$\angle PAB$. Using the number 180 in the postulate is not necessary. As an alternative, we frequently use the irrational number π, which is approximately equal to 3.1416. When 180 is used, the fundamental unit of measure is a **degree**(°). When π is used, the fundamental unit of measure is a **radian** (rad). Thus, for any angle in a plane, the measure of the angle is greater than 0° (or 0 rad) and less than or equal to 180° (or π rad). A restatement of the definition of a straight angle is:

The measure of a **straight angle** is 180° (or π rad).

One degree equals 60 minutes (') and one minute equals 60 seconds ("). Thus, given $\angle ABC$, we may have m$\angle ABC$ = $37°45'36''$. Given $\angle DEF$, we may have m$\angle DEF$ = 1.2 rad. Notice that $0° < 37°45'36'' \le 180°$ and that 0 rad < 1.2 rad $\le \pi$ rad, so that it is possible to have angles with these measures.

Example 1

Perform the indicated operations and simplify.

(a) $47° \ 29' \ 56'' + 59° \ 36' \ 19''$ (b) $172° \ 23' - 89° \ 46' \ 50''$

Solution

(a) $56'' + 19'' = 75'' = 1' \ 15''$ (carry 1')
     $29' + 36' + 1' = 66' = 1° \ 6'$ (carry 1°)
     $47° + 59° + 1° = 107°$
     The answer is 107° 6' 15".

(b) $172° \ 23' = 171° \ 82' \ 60''$ (Borrow 1° and 1')
     $60'' - 50'' = 10''$
     $82' - 46' = 36'$
     $171° - 89° = 82°$
     The answer is 82° 36' 10".

Example 2

If π radians is equivalent to $180°$, how many degrees represent each of the following?

(a) $\pi/3$ rad   (b) $5\pi/6$ rad

Solution

(a) $\pi/3$ rad $= (180/3)° = 60°$
(b) $5\pi/6$ rad $= 5(180/6)° = 5(30)° = 150°$

When finding the length of a line segment, we use a ruler. When finding the measure of an angle, we use a protractor. To find the measure of $\angle BAC$, place point P of the protractor on the vertex A of the angle so that $\overrightarrow{PQ}$ lies on $\overrightarrow{AC}$. Find the intersection of $\overrightarrow{AB}$ with the scale marks on the protractor. The number at that location is the measure of $\angle BAC$ in degrees. See Figure 1.14.

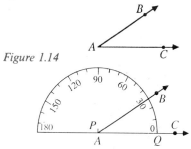

Figure 1.14

Definition

Two angles $\angle ABC$ and $\angle DEF$ are **congruent** if m$\angle ABC =$ m $\angle DEF$. We write $\angle ABC \simeq \angle DEF$.

Notice that $\angle ABC = \angle DEF$ has a different meaning than $\angle ABC \simeq \angle DEF$. Why? (Hint: See the definition of congruent segments in Section 1.3.)

EXERCISE SET 1.4

In Exercises 1–4, perform the indicated operations and simplify.

1. $28°\ 11' + 52°\ 40'$

2. $70°\ 18'\ 32'' + 25°\ 25'\ 40''$

3. $108°\ 45'\ 22'' - 46°\ 32'\ 10''$

4. $165°\ 10'\ 12'' - 74°\ 19'\ 20''$

In Exercises 5–10, use a protractor to draw each of the angles having the following measures.

5. 80°   **8.** 117°

6. 25°   **9.** $\pi/4$ rad

7. 90°   **10.** $2\pi/9$ rad

In Exercises 11–14, if π radians is equivalent to 180°, how many degrees represent:

11. $\pi/2$ rad   **13.** $\pi/8$ rad

12. $3\pi/4$ rad   **14.** $5\pi/12$ rad

15. Discuss the difference between $\angle ABC = \angle DEF$ and $\angle ABC \simeq \angle DEF$.

In Exercises 16–21, give approximate angle measures for each of the events.

16. A telephone pole falling

17. Bending your arm at the elbow as far as possible

18. The turning of a clock's hour hand in 1 hour

19. The turning of a clock's minute hand in 1 minute

20. Turning your head as far as possible

21. Twisting your wrist as far as possible

In Exercises 22–23, assume that a map is being used in which north is toward the top of the map.

22. Draw an angle showing a plane's flight 40° east of north.

23. Draw an angle showing a ship sailing 35° west of south.

1.5 ANGLE AND LINE RELATIONSHIPS

In this section, we discuss some relationships for angles and for lines.

Postulate
1.9

> **(Angle Addition Postulate)** If $D \in \text{int}\angle BAC$, then $\text{m}\angle BAD + \text{m}\angle DAC = \text{m}\angle BAC$.

Figure 1.15

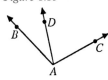

Example 1

In Figure 1.15, if $m\angle BAD = 36°\ 5'$ and $m\angle DAC = 75°\ 58'$, find $m\angle BAC$.

Solution

$$m\angle BAC = 111°\ 63' = 112°\ 3'$$

Definition

Two angles are **complementary** if the sum of their measures is 90°.
Two angles are **supplementary** if the sum of their measures is 180°.

If two angles are complementary, we say that one angle is the **complement** of the other. If two angles are supplementary, we say that one angle is the **supplement** of the other.

Example 2

Find the complement of a 37° angle.

Solution

$90° - 37° = 53°$; thus, $37° + 53° = 90°$.
The answer is 53°.

Example 3

Find the supplement of a 152° 17′ angle.

Solution

$180° - 152°\ 17' = 27°\ 43'$; thus, $152°\ 17' + 27°\ 43' = 180°$.
The answer is 27° 43′.

Example 4

Find the measure of the angle whose supplement is four times its complement.

Solution

Let x be the measure of the angle.
Then $90° - x$ is the measure of its complement,
and $180° - x$ is the measure of its supplement.

We are given that
   supplement = four times complement.

Thus,
$$180° - x = 4 (90° - x).$$

It follows that
$$180° - x = 360° - 4x$$
$$3x = 180°$$
$$x = 60°.$$

Definition

An angle is **acute** if its measure is less than 90°. An angle is **obtuse** if its measure is greater than 90° and less than 180°. An angle is a **right angle** if its measure is 90°.

Example 5

Given $m\angle 1 = \pi/5$ rad, $m\angle 2 = 117°13'$, $m\angle 3 = 90°$, and $m\angle 4 = \pi$ rad, classify each of the angles.

Solution

$\angle 1$ is an acute angle
$\angle 2$ is an obtuse angle.
$\angle 3$ is a right angle.
$\angle 4$ is a straight angle.

Figure 1.16

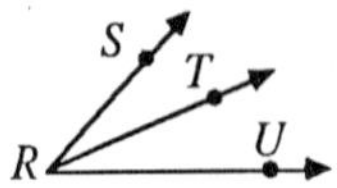

Two angles, $\angle EFG$ and $\angle HFG$ are **adjacent** if G $\in$ int$\angle EFH$. In Figure 1.16, notice that $\angle SRT$ and $\angle URT$ are adjacent, but $\angle SRU$ and $\angle TRU$ are not adjacent. Why?

The above relationship between angles is used to define the following relationships between lines.

Definition

Two lines are **perpendicular** if they intersect so as to form congruent adjacent angles.

Figure 1.17

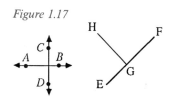

The congruent adjacent angles in the definition of *perpendicular* must be right angles, since the sum of the measures of the two congruent angles is 180°. We write $\overleftrightarrow{AB} \perp \overleftrightarrow{CD}$ if $\overleftrightarrow{AB}$ is perpendicular to $\overleftrightarrow{CD}$. Two line segments are perpendicular if they intersect and are contained in two perpendicular lines. In Figure 1.17, $\overleftrightarrow{AB} \perp \overleftrightarrow{CD}$ and $\overline{EF} \perp \overline{GH}$.

Definition

Consider $\overleftrightarrow{AB}$ and a point C not on $\overleftrightarrow{AB}$. If D is a point of $\overleftrightarrow{AB}$ such that $\overleftrightarrow{AB} \perp \overleftrightarrow{CD}$, then CD is the **distance** from point C to the line $\overleftrightarrow{AB}$.

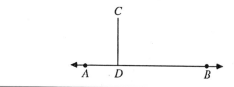

Figure 1.18

Two lines j and k are **equidistant** if the distance from arbitrary points of j to line k is always identical. Thus, in Figure 1.18, if j and k are equidistant, then $AB = CD = EF$. The symbol $\square$ in the figure means *perpendicular*. Two line segments are equidistant if the lines containing them are equidistant. Thus, in Figure 1.18, $\overline{AE}$ and $\overline{BF}$ are equidistant.

EXERCISE SET 1.5

Find the complement of each of the angles in Exercises 1–8.

1. 30°
2. 45°
3. 66°22′
4. 89°10′
5. 3′59″
6. $\pi/6$ rad
7. $\pi/4$ rad
8. $5\pi/18$ rad

Find the supplement of each of the angles in Exercises 9–16.

9. 150°
10. 90°
11. 35°
12. 162°45′
13. 75°35′18″
14. $5\pi/6$ rad
15. $\pi/2$ rad
16. $11\pi/18$ rad

17. What angle is equal in measure to its complement? Its supplement?

18. What are the measures of the angles formed by perpendicular lines?

19. Find the measure of the angle whose supplement is three times its complement.

Figure for Exercises 20–22

Use the figure for Exercises 20–22 to answer the following:

20. List all pairs of adjacent angles.

21. Why are $\angle BAF$ and $\angle EAF$ *not* adjacent?

22. Why are $\angle DBA$ and $\angle BAE$ *not* adjacent?

In Exercises 23–34, classify each angle as acute, obtuse, right, or straight.

23. $120°$

24. $18°$

25. $(32°15' + 57°45')$

26. $(73°18'4'' + 16°41'10'')$

27. $9\pi/20$ rad

28. $3\pi/4$ rad

29. $\pi/2$ rad

30. $5\pi/12$ rad $+ 7\pi/12$ rad

31. The complement of the angle whose measure is $36°$.

32. The supplement of any obtuse angle.

33. The supplement of a straight angle.

34. The complement of the supplement of $165°$.

In Exercises 35–36, give five examples of sports in which the following concepts are important to the playing field (for example, the goal lines in football are equidistant line segments).

35. Equidistant line segments.

36. Perpendicular line segments.

1.6 ELEMENTARY CONSTRUCTIONS

So far, we have been concerned with finding the lengths of line segments and the measures of angles. To do this, we needed a ruler with marks on it and we needed a protractor. In this section, we will discuss geometric constructions. In a **construction**, we may use *only* a straightedge and a compass. A **straightedge** is a ruler with no marks; a **compass** is an instrument for drawing circles.

Construction 1.1

Construct a line segment equal in length to a given line segment.

Given: $\overline{AB}$

Construct: $\overline{CD}$ such that $CD = AB$

A ——————— B

Step 1. Draw a line $\overleftrightarrow{CE}$.

Step 2. Place the points of the compass so that one point is on A and the other point is on B.

Step 3. Without changing the distance between the points of the compass (spread), place one point of the compass on C and the other on $\overrightarrow{CE}$. Label this other point D. Then $\overline{CD}$ is the required segment.

Construction 1.2	Construct an angle equal in measure to a given angle.

Given: $\angle A$

Construct: $\angle B$ such that $m\angle B = m\ A$

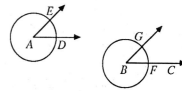

Step 1. Draw a ray $\overrightarrow{BC}$.

Step 2. Choose any convenient spread of the compass and draw a circle A. Notice that $\angle A \cap$ circle A is a two-point set. Label these points D and E.

Step 3. Without changing the spread of the compass, draw a circle B. Let $\overrightarrow{BC} \cap$ circle $B = \{F\}$.

Step 4. Place the points of the compass so that one point is on D and the other point is on E.

Step 5. Without changing the spread of the compass, place one point of the compass on F and the other on circle B. Label this other point G. Draw $\overrightarrow{BG}$. Then $\angle B$ is the required angle.

Definition	Point C is the **midpoint** of $\overline{AB}$ if $A-C-B$ and $AC = BC$.

Definition	A **bisector** of $\overline{AB}$ is any line or line segment that contains the midpoint C, but no other point of $\overline{AB}$.

To make constructions a little clearer, we sometimes avoid drawing complete circles and draw only the parts of the circles that are needed. We do this in the following constructions.

Construction
1.3

> Construct the perpendicular bisector of a given line segment.

Given: $\overline{AB}$

Construct: the perpendicular bisector $\overleftrightarrow{CD}$ of $\overline{AB}$

Step 1. Choose any convenient spread of the compass so that the points of the compass are greater than $AB/2$ apart. Draw circles A and B using this spread of the compass.

Step 2. Circle $A \cap$ circle B is a two-point set. Label these points C and D. Draw $\overleftrightarrow{CD}$. Then $\overleftrightarrow{CD}$ is the required line.

Definition

> The point set $\overrightarrow{AD}$ is the **bisector** of $\angle BAC$ if $D \in \text{int} \angle BAC$ and $\text{m}\angle BAD = \text{m}\angle CAD$.

Construction
1.4

> Construct the bisector of a given angle.

Given: $\angle BAC$

Construct: bisector $\overrightarrow{AD}$ of $\angle BAC$

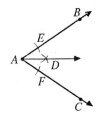

Step 1. Choose any convenient spread of the compass and draw circle A. Notice that $<A \cap$ circle A is a two-point set. Label these points E and F.

Step 2. Choose any convenient spread of the compass so that the points of the compass are greater than $EF/2$ apart. Draw circles E and F using this spread of the compass.

Step 3. Circle $E \cap$ circle F is a two-point set. Label the point D in $\text{int} \angle BAC$. Then $\overrightarrow{AD}$ is the required ray.

Take a moment at this time to consider whether a construction is possible in which the measure of an angle is divided into three equal portions instead of two equal portions.

Unfortunately, it has been proved that the trisection of an arbitrary angle by construction is impossible. Mathematicians attempted to solve this problem for centuries before this disappointing conclusion was reached. Of course, some special angles (such as a straight angle) can be trisected, but the techniques used cannot be generalized.

Construction
1.5

Construct the perpendicular to a line from a point on the line.

Given: $\overleftrightarrow{AB}$

Construct: $\overleftrightarrow{AE} \perp \overleftrightarrow{AB}$

Step 1. Choose any convenient spread of the compass and draw a circle A. Circle $A \cap \overleftrightarrow{AB}$ is a two-point set. Label these points C and D.

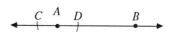

Step 2. Construct the perpendicular bisector $\overleftrightarrow{AE}$ of $\overline{CD}$ (See Construction 1.3). Then $\overleftrightarrow{AE}$ is the required line.

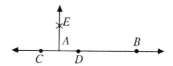

Construction
1.6

Construct the perpendicular to a line from a point not on the line.

Given: $\overleftrightarrow{AB}$, point C

Construct: $\overleftrightarrow{CG} \perp \overleftrightarrow{AB}$

Step 1. Choose any convenient spread of the compass and draw circle C so that circle C intersects $\overleftrightarrow{AB}$ in a two-point set. Label these points D and E.

Step 2. Construct the perpendicular bisector $\overleftrightarrow{FG}$ of $\overline{DE}$. Point $C \in \overleftrightarrow{FG}$, so that $\overleftrightarrow{FG} = \overleftrightarrow{CG}$. Then $\overleftrightarrow{CG}$ is the required line.

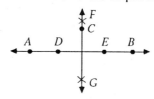

EXERCISE SET 1.6

1. Draw a line segment $\overline{AB}$ with $AB = 2$ inches and draw a line k. Construct $\overline{CD}$ on k such that $\overline{CD} \simeq \overline{AB}$.

2. Given a line segment $\overline{AB}$ 1.5 inches long, construct $\overline{EF}$ such that $EF = 3$ inches.

3. Draw an acute angle and construct an angle of equal measure.

4. Draw an obtuse angle and construct a congruent angle.

5. Using a protractor, draw an angle whose measure is $25°$. Construct an angle of $50°$; an angle of $75°$.

6. Draw a line segment and construct a perpendicular to it.

7. Draw a line segment 3 inches long. By construction, divide it into four equal segments.

8. Draw an acute angle and construct its bisector.

9. Draw an obtuse angle and contruct its bisector.

10. Draw an obtuse angle with measure $130°$. By construction, divide it into four angles of equal measure.

11. Draw a line k and construct a line perpendicular to it at some point on the line.

12. Draw a line segment 2.5 inches long. Construct the perpendicular bisector of the segment.

13. Draw a line k and a point P not on k. Construct the perpendicular from P to k.

14. Draw a straight angle PQR. At Q construct a right angle. Then construct an angle whose measure is $\pi/4$ and an angle whose measure is $\pi/8$.

15. Draw a line k. At any point A on k construct a line $m \perp k$. Then, at point B on m, $B \neq A$, construct a line $n \perp m$. (Lines k and n are called parallel lines.)

16. Go to the library and find information pertaining to the trisection of an angle.

Chapter 1
Key Terms

absolute value, **10**

acute angle, **22**

adjacent angles, **22**

angle, **14**

axioms, **9**

between, **11**

bisector of a line segment, **25**

bisector of an angle, **26**

center of a circle, **11**

circle, **11**

closed half-plane, **17**

compass, **24**

complement, **21**

complementary angles, **21**

congruent angles, **19**

congruent line segments, **13**

construction, **24**

coordinate, **10**

degree, **18**

distance between two points, **11**

distance from a point to a line, **23**

edge of a half-plane, **17**

elements, **2**

empty set, **4**

endpoints, **13**

equal sets, **3**

equidistant, **23**

exterior angle of a triangle, **16**

exterior of an angle, **15**

exterior of a triangle, **16**

finite, **3**

half-plane, **17**

infinite, **3**

interior angle of a triangle, **16**

interior of an angle, **15**

interior of a triangle, **15**

intersection, **4**

length of a line segment, **13**

line, **2**

line segment, **13**

logical system, **2**

measure of an angle, **18**

midpoint, **25**

minutes, **18**

obtuse angle, **22**

on opposite sides of a line, **14**

on the same side of a line, **14**

one-to-one correspondence, **3**

perpendicular, **22**

plane, **2**

point, **2**

postulate, **9**

radian, **18**

radius (radii) of a circle, **11**

ray, **14**

remote interior angles, **16**

right angle, **22**

seconds, **18**

set, **2**

side of an angle, **14**

side of a triangle, **15**

side opposite an angle, **16**

space, **9**

straight angle, **14**

straightedge, **24**

subset, **4**

supplement, **21**

supplementary angles, **21**

theorems, **10**

triangle, **15**

union, **4**

vertex of an angle, **14**

Chapter 1
Review Exercises

Fill each blank with the word *always, sometimes,* or *never.*

1. The empty set is _____ a subset of any set S.

2. Zero is _____ an element of the empty set.

3. If A and B are sets, then $A \cap B$ is _____ the empty set.

4. If A and B are sets, $x \in B$, and $x \notin A$, then $A \cup B$ is _____ the set A.

5. Three distinct points in space are _____ in a plane.

6. The absolute value of a real number is _____ positive.

7. If $A{-}B{-}C$ and $A{-}B{-}D$, then $A{-}C{-}D$ is _____ true.

8. If $\overline{AB} \simeq \overline{PQ}$, then $\overline{AB} = \overline{PQ}$ is _____ true.

9. An exterior angle of a triangle is _____ obtuse.

10. Complementary angles are _____ adjacent angles.

True-false: If the statement is true, mark it so. If it is false, replace the underlined word to make a true statement.

11. The set of squares of the even integers is <u>finite</u>.

12. If $A \cap B = \emptyset$ and $B \neq \emptyset$, then $A \cup B$ contains <u>at least one</u> element not in A.

13. If $A = \{1, 2, 3, 4, 5\}$ and $B = \{2, 4, 6, 8\}$, then $\{2, 4\}$ is the <u>union</u> of A and B.

14. Any <u>two</u> distinct points in space have exactly one line that contains them.

15. The absolute value of $x - y$ is <u>$y - x$</u> if $y > x$.

16. If $A{-}B{-}C$ and $C{-}D{-}E$, then $A{-}B{-}D$.

17. Point $\underline{A}$ is the endpoint of $\overrightarrow{BA}$.

18. If P and Q are points on the same side of line k, and P and R are on opposite sides of k, then Q and R are <u>on the same side</u> of k.

19. The <u>complement</u> of an obtuse angle is an acute angle.

20. If $\angle ACB$ is a straight angle and $\overleftrightarrow{CD}$ bisects $\angle ACB$, then $\overleftrightarrow{CD}$ <u>is perpendicular</u> to $\overleftrightarrow{AB}$.

In Exercises 21–27, state whether each of the sets is finite, infinite, or the empty set.

21. The points C such that $A-C-B$

22. The values of $|x|$ if $x \in \{-3, -2, -1, 0, 1, 2, 3\}$

23. The sides of a given triangle

24. The exterior angles of a triangle

25. The real numbers x such that $|x| < 0$

26. The points $P \in \text{int} \angle ABC$

27. The obtuse angles DEF such that $m\angle DEF = 32°$

In Exercises 28–35, let $A = \{1, 2, 3, 4, 5\}$, $B = \{1, 4, 9, 16, 25\}$, $C = \{a, b, c, d, e\}$, and $D = \{4\}$.

28. Which sets are in one-to-one correspondence?

29. The set D is a subset of what set or sets?

30. Find $A \cup B$

31. Find $C \cup D$

32. Find $A \cap D$

33. Find $D \cap \emptyset$

34. Find $\emptyset \cup C$

35. Find $\emptyset \cup (A \cap B)$

In Exercises 36–41, find the absolute values.

36. -2

37. $-(-3)$

38. x, if $x < 0$

39. x^2 if $x < 0$

40. 0

41. $a - b$, if $b > a$

42. List all of the numbered angles that are exterior angles of △*ABC*.

43. List the pairs of remote interior angles for each exterior angle listed in Exercise 42.

44. List all pairs of the numbered angles that are adjacent angles.

45. Which of the pairs of adjacent angles in Exercise 44 are supplementary?

46. If ∠3 is acute, is ∠7 acute, right, or obtuse? Why?

Figure for Exercises 42-46

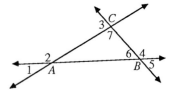

Find the complement of each of the angles in Exercises 47–50, if possible.

47. 80°

48. 43° 28′

49. 2π/3 rad

50. 50″

51–54. Find the supplement of each of the angles in Exercises 47–50.

Constructions:

55. Draw a line segment $\overline{AB}$ with $AB = 1.5$ inches and a line k. Construct $\overline{CD}$ on k such that $\overline{CD} \simeq \overline{AB}$. Construct $\overline{CE}$ on k such that $CE = AB/2$ and $\overline{CF}$ such that $CF = 2AB$.

56. Draw a line k. Construct a perpendicular to k at point P on k. Then construct a 45° angle with P as a vertex.

57. Draw a line k and a point Q not on k. Construct the perpendicular from Q to k.

58. Draw a fairly large triangle and construct the bisectors of its three angles. (They should intersect in a single point.)

59. Draw a fairly large triangle and construct the perpendicular bisectors of the three sides. (They should intersect in a single point.)

Proofs and Congruent Triangles

David Hilbert (1862–1943)

Historical Note

Euclid, who lived about 300 B.C., greatly enhanced the understanding of mathematics by collecting and organizing all the geometric facts known at the time. He published thirteen books called [the] *Elements*. Many of the ideas presented in these books had been known long before, and Euclid's teacher, Thales, may have been the first developer of demonstrative geometry. Some of the logic in the study of geometry was significantly modified centuries later by mathematicians such as David Hilbert (1862-1943).

In our discussion of a logical system, we described undefined terms, definitions, and postulates as they are applied to plane geometry. In this chapter, we begin the discussion of theorems, which are statements that can be proved in terms of the elements of a logical system.

2.1 NEGATION, CONJUNCTION, AND DISJUNCTION (Optional)

Here, we discuss the precise relationship among the English words *not*, *and*, and *or* within a statement and the truth or falsity of the statement.

Definition | If "p" is a statement, the statement "not p" is the **negation** of "p."

In ordinary conversational English, the *not* in a sentence is often placed in a location other than the beginning.

Example 1

Form the negation of the following statements:
(a) The moon is rising.
(b) $\angle ABC$ is a remote interior angle.
(c) Point C is between points A and B.
(d) $m\angle 3 = 25°$.

Solution

(a) The moon is not rising.
(b) $\angle ABC$ is not a remote interior angle.
(c) Point C is not between points A and B.
(d) $m\angle 3 \neq 25°$.

∎

If we know whether a statement is true or false, then a precise rule exists for determining whether the negation of the statement is true or false. A rule of this type can be demonstrated through the use of a **truth table**, such as the one following.

p	$not\ p$
T	F
F	T

The truth table indicates that if the statement "p" is true, then the statement "not p" is false and if the statement "p" is false, then the statement "not p" is true.

Example 2

Indicate whether the following statements and their negations are true or false.

(a) $9 = 2 + 7$.
(b) The moon is in the Atlantic Ocean.
(c) There is a one-to-one correspondence between the set of all real numbers and the set of all points on a line.

Solution

(a) The statement "$9 = 2 + 7$" is true; thus, the statement "$9 \neq 2 + 7$" is false.
(b) The statement "the moon is in the Atlantic Ocean" is false; thus, the statement "the moon is not in the Atlantic Ocean" is true.
(c) The statement "there is a one-to-one correspondence between the set of all real numbers and the set of all points on a line" is true by the Ruler Postulate; thus, the statement "there is not a one-to-one correspondence between the set of all real numbers and the set of all points on a line" is false.

Definition

> If "p" and "q" are statements, the compound statement "p and q" is the **conjunction** of "p" and "q".

Example 3

Form the conjunction of the following pairs of statements.

(a) I am hungry. The dinner is ready.
(b) Points A and B are on a line. Points A and C are on a line.
(c) $m\angle ABC = 30°$. $m\angle CAB \neq 45°$.

Solution

(a) I am hungry and the dinner is ready.
(b) Points A and B are on a line and points A and C are on a line.
(c) m$\angle ABC$ = 30° and m$\angle CAB$ ≠ 45°.

■

If one knows whether two statements are true or false, then a precise rule exists for determining whether the conjunction of the two statements is true or false. The truth table below indicates the relationship of two statements and their conjunction.

p	q	p and q
T	T	T
T	F	F
F	T	F
F	F	F

Thus, if statement "p" is true and statement "q" is true, then the statement "p and q" is true. In all other possible cases, the statement "p and q" is false.

■

Example 4

Indicate whether the following statements and their conjunctions are true or false.

(a) Two angles are complementary if the sum of their measures is 90°. Two angles are supplementary if the sum of their measures is 180°.
(b) An angle is acute if its measure is greater than 90°. An angle is a right angle if its measure is 90°.
(c) 6 ≠ 4 + 3. The radius of all circles is 5 feet.
(d) Every line contains, at most, one point. Every plane contains all points of space.

Solution

(a) Both statements are true; thus, their conjunction is true.
(b) The first statement is false. The second statement is true. Their conjunction is false.
(c) The first statement is true. The second statement is false. Their conjunction is false.
(d) Both statements are false; thus, their conjunction is false.

■

Definition	If "p" and "q" are statements, the compound statement "p or q" is the **disjunction** of "p" and "q".

■

Example 5

Form the disjunction of the pairs of statements in Example 3.

Solution

(a) I am hungry or the dinner is ready.
(b) Points A and B are on a line or points A and C are on a line.
(c) $m\angle ABC = 30°$ or $m\angle CAB \neq 45°$.

■

The truth table below indicates the relationship between two statements and their disjunction.

p	q	p or q
T	T	T
T	F	T
F	T	T
F	F	F

Thus, the disjunction of two statements is true if either one or both of the statements are true. The only case in which a disjunction is false is if both statements are false.

■

Example 6

Indicate whether the statements in Example 4 and their disjunctions are true or false.

Solution

(a) Both statements are true; thus, their disjunction is true.
(b) The first statement is false. The second statement is true. Their disjunction is true.
(c) The first statement is true. The second statement is false. Their disjunction is true.
(d) Both statements are false; thus, their disjunction is false.

■

It is possible to produce more complex sentences by combining the words *not, and,* and *or* in one statement. Truth tables are very useful in determining truth or falsity of complex statements.

Example 7

Make a truth table for the statement "not p or q."

Solution

p	*q*	*not p*	*not p or q*
T	T	F	T
T	F	F	F
F	T	T	T
F	F	T	T

In Example 7, the "not" in front of the p only refers to the p. If we wish the "not" to refer to more than one simple statement, then we must use a grouping symbol. For example, we might consider the statement "not (p or q)," which has a different meaning and truth table than "not p or q."

Example 8

Make a truth table for the statement "p or (r and not q)."

Solution

p	*q*	*r*	*not q*	*(r and not q)*	*p or (r and not q)*
T	T	T	F	F	T
T	T	F	F	F	T
T	F	T	T	T	T
T	F	F	T	F	T
F	T	T	F	F	F
F	T	F	F	F	F
F	F	T	T	T	T
F	F	F	T	F	F

We see from the solution of Example 8 that in five of eight possible situations, the statement "p or (r and not q)" is true and in three situations, the statement is false.

EXERCISE SET 2.1

Form the negation of each of the statements in Exercises 1–10.

1.   I will earn enough money.

2.   I will buy a car.

3.   You will not turn the television off.

4.   You will not be able to study better.

5.   Two angles are not complementary.

6.   Two angles are adjacent.

7. $\overline{AB} \simeq \overline{CD}$.

8. $m\angle A = m\angle B$.

9.   A point is in the interior of $\triangle PQR$.

10.   A point is in $\text{int}\angle Q$.

11–15.   If the statements in Exercises 1–5 are true, are the negations of the statements true or false?

16–20.   If the statements in Exercises 6–10 are false, are the negations of the statements true or false?

Form the conjunction of each of the pairs of statements in Exercises 21–25.

21.   I will earn enough money. I will buy a car.

22.   You will not turn the television off. You will not be able to study better.

23.   Two angles are not complementary. Two angles are adjacent.

24. $\overline{AB} \simeq \overline{CD}$. $m\angle A = m\angle B$.

25.   A point is in the interior of $\triangle PQR$. A point is in $\text{int}\angle Q$.

26–30.   Form the disjunction of each of the pairs of statements in Exercises 21–25.

Indicate whether the conjunctions of the pairs of statements in Exercises 31–35 are true or false.

31.   A logical system consists of undefined terms, definitions, assumptions, and theorems. A set is finite if the number of elements in the set is equal to some nonnegative integer.

32. A set is empty if zero is its only element. The set A is a subset of the set B if every element in A is also an element of B.

33. Every plane contains at least three distinct points, not all on one line. Every plane contains all possible lines in space.

34. $|x| = x$ for all real numbers x. If x and y are coordinates of points P and Q on a line, then the distance from P to Q is $|x + y|$.

35. If $B-A-C$, then $\angle BAC$ is a straight angle. A half-plane is the union of all points on one side of a line.

36–40. Indicate whether the disjunctions of the pairs of statements in Exercises 31–35 are true or false.

41. Give an example of a *not* statement that you might hear in a sports event. Write the negation of the statement.

42. Give an example of two statements that you might hear in a supermarket. Write the conjunction and disjunction of the statements.

43. Give an example of two statements that you might hear during lunch. Write the conjunction and disjunction of the statements.

44. Give an example of two statements that you might hear on a date. Write the conjunction and disjunction of the statements.

45. Make a truth table for the statement "p or not p."

46. Make a truth table for the statement "p and not p."

47. Make a truth table for the statement "p and (r or not q)."

48. Make a truth table for the statement "p or (r or not q)."

49. Make a truth table for the statement "not p and (p or r)."

50. Make a truth table for the statement "not p or (p and r)."

In Exercises 51–52, we show statements of De Morgan's laws. Use truth tables to show that the laws are correct.

51. not (p or q) = not p and not q

52. not (p and q) = not p or not q

If set S is a subset of U, then the set S' of elements that are in U, but not in S, is called the complement of S. Show that De Mor-

gan's laws are true in each of the following Exercises 53-54, when $S = \{2, 4, 8\}$, $T = \{2, 10\}$, and $U = \{2, 4, 6, 8, 10\}$.

53. $(S \cup T)' = S' \cap T'$ *S' = complement of   S = the set of All elements in U But Not in S*

54. $(S \cap T)' = S' \cup T'$

Show that De Morgan's laws are true in each of the following Exercises 55–56, when $S = \{$the set of positive integers$\}$, $T = \{$the set of negative integers$\}$, and $U = \{$the set of all integers$\}$.

55. See No. 53.

56. See No. 54.

2.2   HYPOTHESIS AND CONCLUSION

Theorems are often stated in "if-then" form. In general, given two statements "p" and "q", a statement is in "if-then" form if it reads "if p, then q." The statement "p," written as a complete sentence independent of any other sentence, is the **hypothesis**. The statement "q," written as a complete sentence independent of any other sentence, is the **conclusion**. The word *if* serves as an introductory symbol indicating that the hypothesis follows, and *then* indicates that the conclusion follows.

Example 1

Give an example of a statement in "if-then" form. Give the hypothesis and conclusion of the statement.

Solution

The statement "If it is cloudy, then it will rain," is in "if-then" form. The hypothesis is "it is cloudy" and the conclusion is "it will rain."

Science textbooks contain many statements with hypotheses and conclusions. The statement in the following example might come from a textbook on genetics.

Example 2

Find the hypothesis and conclusion of the statement "If it is a matter of chance which genotype mates with which other genotype, then it is a matter of chance which allele combines with which other allele."

Solution

The hypothesis is "It is a matter of chance which genotype mates with which other genotype" and the conclusion is "It is a matter of chance which allele combines with which other allele."

In the English language, the word *then* is frequently omitted.

Example 3

Find the hypothesis and conclusion of the statement "If I eat too much candy, I will get sick."

Solution

The word *then* has been left out after the comma; however, the sentence is to be treated as if the *then* were there. Hence, the hypothesis is "I eat too much candy" and the conclusion is "I will get sick."

At other times in the English language, the order of the *if* clause and the *then* clause are reversed.

Example 4

Find the hypothesis and conclusion of the statement "I will be very upset, if I fail geometry."

Solution

The hypothesis is "I fail geometry" and the conclusion is "I will be very upset."

These examples illustrate that a person must inspect a statement very closely in order to determine which part of it is the hypothesis and which part is the conclusion; however, the word *if* serves as an important clue. Unfortunately, in many English statements, the word *if* is omitted.

Example 5

Find the hypothesis and conclusion of the statement "All crazy people like porcupines."

Solution

Before searching for the hypothesis or conclusion, one should translate this sentence into "if-then" form, namely "If a person is crazy, then he likes porcupines." The hypothesis is then clearly "a person is crazy," and the conclusion is "a person likes porcupines." In order to make the second sentence independent of the first, the pronoun "he" was replaced by the noun phrase it represented.

Suppose we are given the two statements "if p, then q," and "if q, then p." The hypothesis and conclusion of the first sentence have been interchanged to form the second sentence. Such statements are **converses** of each other. This is not the idea that was illustrated in Example 3. In that example, the hypothesis and conclusion were not changed, but they were written in a different location in the sentence.

Example 6

Find the converse of the statement "If A, B, and C are on the same line and $AB + BC = AC$, then B is between A and C."

Solution

The converse is "If B is between A and C, then A, B, and C are on the same line and $AB + BC = AC$."

In Example 6, the sentence given is the definition of "between." In this case, the statement and its converse are both true. This is a requirement in a definition, as stated in the following principle:

> A definition must be reversible; that is, the statement and its converse must both be true.

By this principle, if we give a definition, we know that it is reversible. Thus, it is redundant to give both the statement and a converse within a *definition*, and most mathematical journals consider it poor form to do so. However, if a *theorem* is given, and both a statement and its converse are true, then this must be explicitly stated. In general, a statement and its converse need not both be true.

Given the statement "If p, then q," its **inverse** is "If not p, then not q." The converse of the inverse is the **contrapositive** of the original statement. Thus, given the statement "If p, then q," its contrapositive is "If not q, then not p."

Example 7

Find the converse, inverse, and contrapositive of the statement "Two angles are complementary if the sum of their measures is 90°."

Solution

In "if-then" form, the statement is "If the sum of the measures of two angles is 90°, then the angles are complementary." The converse is "If two angles are complementary, then the sum of their measures is 90°." The inverse is "If the sum of the measures of two angles is not 90°, then the angles are not complementary." The contrapositive is "If two angles are not complementary, then the sum of their measures is not 90°."

The original statement in Example 7 was given earlier as a definition—the converse, inverse, and contrapositive of a definition are always true, and the contrapositive of a theorem is always true. In general, if a statement is true, then its contrapositive is true, and if a statement is false, then its contrapositive is false. Notice that the inverse of a statement is the contrapositive of the converse of a statement. Thus, if the inverse of a statement is true, then the converse of the statement is true, and if the inverse of a statement is false, then the converse of the statement is false.

■

Example 8

Given the statement "If it is cloudy, then it is raining," determine whether it is true or false. Do the same for its contrapositive, its converse, and its inverse.

Solution

The statement is false, as is its contrapositive "If it is not raining, then it is not cloudy." However, the converse "If it is raining, then it is cloudy" is true, as is its inverse "If it is not cloudy, then it is not raining."

■

Figure 2.1

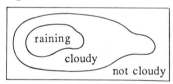

Figure 2.2

Figure 2.3

Example 8 can be illustrated by a **Venn diagram**. In Figure 2.1, we see that the set "raining" is subset of the set "cloudy." Thus, if it is raining, then it is cloudy. But there are points of the set "cloudy" that are not in the set "raining." Thus, it is false that if it is cloudy, then it is raining. It is clear from the diagram that if it is not cloudy, then it is not raining, since all the points of the set "raining" are contained in the set "cloudy." Also, from the diagram we see that if it is not raining, it still might be cloudy.

In summary, the four types of statements are listed below.

1. "If p, then q." (original statement)

2. "If q, then p." (converse of original statement)

3. "If not p, then not q." (inverse of original statement)

4. "If not q, then not p." (contrapositive of original statement)

In this summary, Statements 1 and 4 are always logically equivalent and Statements 2 and 3 are always logically equivalent. Statements 1 and 4 are illustrated by the Venn diagram in Figure 2.2. Statements 2 and 3 are illustrated by the Venn diagram in Figure 2.3.

EXERCISE SET 2.2

Write the hypothesis and the conclusion as independent sentences for each of the statements in Exercises 1-10.

1. If I earn enough money, then I will buy a car.

2. If you eat too much, you will get fat.

3. If you turn the television off, you will be able to study better.

4. Two angles are complementary if the sum of their measures is $90°$.

5. If $A \simeq B$, then $m\angle A = m\angle B$.

6. A set that has at least one element cannot be the empty set.

7. "Red in the morning, sailors take warning."

8. Set $C = \{1, 2, 3, 4, 5, 6\}$ is the union of the sets $A = \{1, 3, 5\}$ and $B = \{2, 4, 6\}$.

9. A theorem always has a hypothesis and a conclusion.

10. A point that is in the interior of $\triangle PQR$ is in $\text{int}\angle Q$.

11-20. Write the converse of each of the statements in Exercises 1-10.

21-30. Write the inverse of each of the statements in Exercises 1-10.

31-40. Write the contrapositive of each of the statements in Exercises 1-10.

For each of the statements in Exercises 41–48, find the hypothesis, conclusion, converse, inverse, and contrapositive.

41. On the moon, the acceleration due to gravity is only about one-sixth that on earth.

42. If we ignore the effect of air friction, all bodies fall to earth with the same acceleration.

43. Rubber has the property of becoming sticky when it is warmed.

44. Glycerol and sodium salts of the fatty acids are formed when animal fat is treated with sodium hydroxide.

45. Whenever a reactive metal is made the anode of an electrolytic cell, the anodic oxidation may involve oxidation of the metal composing the electrode.

46. Any particle in suspension will be bombarded on all sides by moving molecules of the dispersion medium.

47. Any zygote formed by combination with a normal gamete will be aneuploid and lethal.

48. An example of sex-influenced inheritance is pattern baldness, in which premature hair loss occurs on the front and top of the head, but not on the sides.

49. A taxpayer who uses the accrual basis reports income when it is earned, even though not yet received, and deducts expenses when they are incurred, even though not yet paid.

50. When a company asset is sold, or otherwise disposed of, it is important to record the depreciation up to the date of sale or disposition.

51. Give an example of an "if-then" statement that you might hear in a sports event. Write the converse, inverse, and contrapositive of the statement.

52. Give an example of an "if-then" statement that you might hear in a supermarket. Write the converse, inverse, and contrapositive of the statement.

53. Give an example of an "if-then" statement that you might hear during lunch. Write the converse, inverse, and contrapositive of the statement.

54. Give an example of an "if-then" statement that you might hear on a date. Write the converse, inverse, and contrapositive of the statement.

2.3 PREPARING FOR A PROOF

After learning how to find the hypothesis and conclusion in a given statement, we are ready to prove the statement. First, translate the statement into a **given** part and a **prove** part. These parts are essentially the hypothesis of the statement and the conclusion of the statement, respectively. Represent these symbolically, in a diagram drawn to illustrate the theorem and help plan the steps in the **proof**. In this section we only set up the proof by drawing a diagram and finding out what is *given* and what we are trying to *prove*. In later sections, we discuss how to complete the proofs.

■

Example 1

Set up the proof for the statement "A bisector of an interior angle of a triangle intersects the side opposite the angle."

Solution

Step 1. Translated into "if-then" form, the statement reads "If an interior angle of a triangle has a bisector, then the bisector intersects the side opposite the angle."

Step 2. The hypothesis is "an interior angle of a triangle has a bisector." The conclusion is "the bisector intersects the side opposite the angle."

Figure 2.4

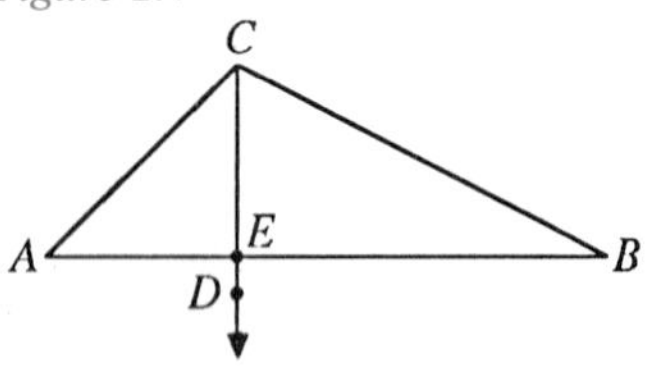

Step 3. A typical drawing representing the preceding statements is shown in Figure 2.4.

Step 4. Translate the statements in Step 2 in terms of the diagram:

Given: $\overrightarrow{CD}$ bisects $\angle ACB$

Prove: There is a point E such that $\overrightarrow{CD} \cap \overline{AB} = \{E\}$

■

■

Example 2

Set up the proof for the statement "If two interior angles of a triangle are congruent, then the sides opposite these angles are congruent."

Solution

Step 1. The hypothesis is "Two interior angles of a triangle are congruent," and the conclusion is "The sides opposite these angles are congruent."

Figure 2.5

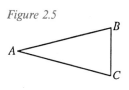

Step 2. A possible drawing representing the previous statement is shown in Figure 2.5.

Step 3. *Given:* $\angle B \simeq \angle C$
Prove: $\overline{AC} \simeq \overline{AB}$

EXERCISE SET 2.3

Make and label a drawing to illustrate each of the following statements, then write what is *given* and what you wish to *prove* in terms of the diagram. (If you need help with axioms, symbols, or abbreviations, see Appendices A, E, or F.)

1. The sum of the measures of the angles of a triangle is 180°.

2. If two sides of a triangle are congruent, the bisector of the angle formed by these two sides is perpendicular to the third side.

3. In a triangle, the square of the length of the side opposite a right angle is equal to the sum of the squares of the lengths of the other two sides.

4. The supplements of congruent angles are congruent.

5. If two lines intersect, the bisectors of a pair of adjacent angles formed are perpendicular.

6. The bisectors of the three angles of a triangle intersect in the same point.

7. The perpendicular bisectors of the three sides of a triangle intersect in the same point.

8. A triangle can have no more than one right angle.

9. The supplement of an obtuse angle is acute.

10. The converse of Exercise 3.

Write what is *given* and what you wish to *prove* in the following statements from algebra.

11. If $x + 7 = 9$, then $x = 2$.

12. If $y - 6 = 4$, then $y = 10$.

13. If $f(x) = x - 9$ and $g(y) = y + 6$, then $f(x) + g(y) = x + y - 3$.

14. If $f(x) = x + 2$ and $g(y) = y - 8$, then $f(x) + g(y) = x + y - 6$.

2.4 PROOFS

In this section are two examples of complete proofs of theorems. A complete proof consists of:

1. The theorem to be proved

2. Diagram

3. Given

4. Prove

5. Proof: (a) **Statements**
           (b) **Reasons** justifying the statements

Between Steps 4 and 5 above, an **analysis** is needed and may be written as a *plan of proof*, though a written analysis is generally not required. However, it is the analysis that determines what is to be written in Step 5. Decide which statements are going to be needed and whether these statements can be justified. A statement must be justified by one of the following:

1. Given statements

2. Definitions

3. Axioms or postulates

4. Previously proved theorems

We are *not* allowed to make up a statement which *seems* to be true and use it as justification.

Theorem 2.1 | Supplements of congruent angles are congruent.

Given: $\angle 1$ and $\angle 2$ are supplementary
$\angle 3$ and $\angle 4$ are supplementary
$\angle 2 \simeq \angle 4$

Prove: $\angle 1 \simeq \angle 3$

Analysis: In developing an analysis, we start with what we are to prove and work backward to what is given. To prove that two angles are congruent, we need to show that their measures are equal. This can be done by using the definition of supplementary angles and by substituting for congruent angles. We then do the necessary algebra, and hope for the best! The actual proof must proceed from what is given to what we need to prove.

Proof:

Statements	Reasons
1. $\angle 1$ and $\angle 2$ are supp. $\angle 3$ and $\angle 4$ are supp.	1. Given
2. $m\angle 1 + m\angle 2 = 180°$ $m\angle 3 + m\angle 4 = 180°$	2. Def. of supp.
3. $m\angle 1 + m\angle 2 = m\angle 3 + m\angle 4$	3. Substitution (or symmetry *and* transitivity)
4. $\angle 2 \simeq \angle 4$	4. Given
5. $m\angle 2 = m\angle 4$	5. Def. of $\simeq \angle$ s
6. $m\angle 1 + m\angle 2 = m\angle 3 + m\angle 2$	6. Statements 3 and 5; substitution
7. $m\angle 2 = m\angle 2$	7. Reflexivity
8. $m\angle 1 = m\angle 3$	8. Statements 6 and 7; Add. Axiom of =
9. $\angle 1 \simeq \angle 3$	9. Def. of $\simeq \angle$ s

Several points to be considered in the above proof are:

(a) Statement 4 could have been included with Statement 1; however, it was not needed until after Statement 3. In general, several orders of statements are possible in a proof, but be careful not to write a statement that requires information not yet included in the proof. For example, do not write Statement 6 before Statement 5 because Statement 6 depends on Statement 5.

(b) Statements 3 and 7 refer to axioms learned in basic algebra, namely reflexivity, symmetry, and transitivity. These and other axioms are reviewed in Appendix A.

(c) Abbreviations and symbols may be used in proofs. Thus *def.* is used for *definition*, *supp.* is used for *supplementary*, and $\angle$ is used for *angle*. We use $\simeq$ for *congruent* even though in general it means *is congruent to*." See the list of abbreviations in Appendix F.

(d) Some statements, such as Statement 3, have several possible reasons.

(e) When a statement involves several other preceding statements, tell what other statements are involved. For instance, in the reason for Statement 6, Statements 3 and 5 are involved. It is not strictly necessary to include this information, but such information helps a person follow the steps in a proof.

The proof of the following theorem is very similar to the proof of the preceding one, and so the proof is left to the reader as an exercise.

Theorem 2.2

> Complements of congruent angles are congruent.

Definition

> If two rays $\overrightarrow{AB}$ and $\overrightarrow{AC}$ are given, such that $C-A-B$, then the rays are **opposite rays**.

Figure 2.6

See Figure 2.6 for an illustration of opposite rays $\overrightarrow{AB}$ and $\overrightarrow{AC}$.

Definition

> If two angles with the same vertex are placed such that the sides of one angle form opposite rays with the sides of the other angle, then the angles are **vertical angles**.

Figure 2.7

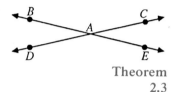

In Figure 2.7, $\angle BAC$ and $\angle DAE$ are vertical angles, and $\angle BAD$ and $\angle CAE$ are also vertical angles.

Theorem 2.3

> Vertical angles are congruent.

Given: $\angle 2$ and $\angle 3$ are vertical angles

Prove: $\angle 2 \simeq \angle 3$

Proof:

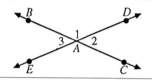

Statements	Reasons
1. $\angle 2$ and $\angle 3$ are vert. $\angle$ s	1. Given
2. $\overrightarrow{AB}$ and $\overrightarrow{AC}$ are opp. rays $\overrightarrow{AD}$ and $\overrightarrow{AE}$ are opp. rays	2. Def. of vert. $\angle$ s
3. $\angle BAC$ and $\angle DAE$ are st. $\angle$ s	3. Def. of opp. rays and st. $\angle$ s
4. $m\angle BAC = 180°$ $m\angle DAE = 180°$	4. Protractor Post. and def. of meas.
5. $m\angle 1 + m\angle 2 = m\angle BAC$ $m\angle 1 + m\angle 3 = m\angle DAE$	5. Angle Add. Post.
6. $m\angle 1 + m\angle 2 = 180°$ $m\angle 1 + m\angle 3 = 180°$	6. Transitivity

7. $m \angle 1 + m \angle 2 = m \angle 1 + m \angle 3$ 7. Substitution

8. $m \angle 2 = m \angle 3$ 8. Add. Axiom of $=$

9. $\angle 2 \simeq \angle 3$ 9. Def. of $\simeq$

EXERCISE SET 2.4

1. List the five steps of a complete proof.

2. List the four types of justifying statements that may be used in a proof.

3. Name all the pairs of vertical angles in the diagram.

Figure for Exercise 3

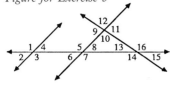

Complete the proof of each of Exercises 4–16.

4. Given: $m \angle BAC = m \angle EDF$
   $\overrightarrow{AG}$ bisects $\angle BAC$
   $\overrightarrow{DH}$ bisects $\angle EDF$

Figure for Exercises 4-16

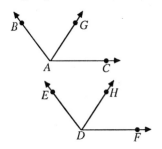

Prove: $m \angle GAC = m \angle HDF$

Proof:

Statements	Reasons
1. ?	1. Given
2. $m \angle BAC = m \angle BAG +$ $m \angle GAC$ $m \angle EDF = m \angle EDH +$ $m \angle HDF$	2. ? Angle Add, tm Postulate Post 1.9
3. $m \angle BAG + m \angle GAC =$ $m \angle EDH + m \angle HDF$	3. ? subst
4. $m \angle BAG = m \angle GAC$ $m \angle EDH = m \angle HDF$	4. ? of Biset
5. $2 m \angle GAC = 2 m \angle HDF$	5. ? Sub
6. $m \angle GAC = m \angle HDF$	6. ?

Figure for Exercise 5

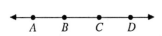

5. Given: $A - B - C$, $B - C - D$,
   $AB = CD$

Prove: $AC = BD$

Proof:

Statements	Reasons
1. ?	1. Given
2. $AC = AB + BC$	2. ?
3. $AC = CD + BC$	3. ?
4. $CD + BC = BC + CD$	4. ?
5. $AC = BC + CD$	5. ?
6. $AC = BD$	6. ?

Figure for Exercise 6

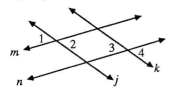

6. *Given:* $j, k, m,$ and n are lines,
   $$m\angle 2 = m\angle 3$$
   Prove: $m\angle 1 = m\angle 4$
   Proof:

Statements	Reasons
1. ?	1. ? Given
2. ? ∂1+∠2 ≅, 3→4	2. Vert. $\angle$s are $\simeq$
3. $m\angle 1 = m\angle 4$	3. ? Trans

Figure for Exercise 7

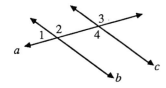

7. *Given:* $a, b,$ and c are lines,
   $$m\angle 1 + m\angle 3 = 180°$$
   Prove: $m\angle 2 = m\angle 4$
   Proof:

Statements	Reasons
1. ?	1. Given
2. $m\angle 1 + m\angle 2 = 180°$	2. ?
3. $m\angle 1 + m\angle 2 = m\angle 1 + m\angle 3$	3. ?
4. ?	4. Add. Axiom of $=$
5. $m\angle 3 = m\angle 4$	5. ?
6. ?	6. Transitivity

8. *Given:* j and k are lines,
   $$m\angle 1 = m\angle 3$$
   Prove: $m\angle ABE = m\angle CBD$

Figure for Exercise 8

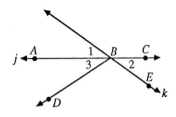

Proof:

Statements	Reasons
1. j and k are lines	1. ?
2. $m\angle 2 = m\angle 1$	2. ?
3. $m\angle 1 = m\angle 3$	3. ?
4. $m\angle 2 = m\angle 3$	4. ?
5. ?	5. Def. of supp.
6. $m\angle ABE = m\angle CBD$	6. ?

Figure for Exercise 9

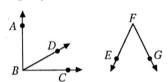

9. *Given:* $\angle ABC$ is a right angle,
        $\angle EFG$ and $\angle DBC$ are complementary
   Prove: $m\angle ABD = m\angle EFG$
   Proof:

Statements	Reasons
1. ?	1. Given
2. $m\angle ABC = 90°$	2. ?
3. $m\angle ABD + m\angle DBC = m\angle ABC$	3. ?
4. ?	4. Transitivity
5. $\angle ABD$ and $\angle DBC$ are comp.	5. ?
6. $m\angle ABD = m\angle EFG$	6. ?

Figure for Exercise 10

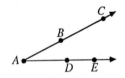

10. *Given:* rays $\overrightarrow{AC}$ and $\overrightarrow{AE}$,
         $AC = AE$, $AB = AD$
    Prove: $BC = DE$
    Proof:

Statements	Reasons
1. $AB + BC = AC$, $AD + DE = AE$	1. ? Def of Bet
2. $AC = AE$	2. ? given
3. $AB + BC = AD + DE$	3. ? sub
4. $AB = AD$	4. ? given
5. $AD + BC = AD + DE$	5. ? sub
6. $BC = DE$	6. ? — subtraction

continue study

Figure for Exercise 11

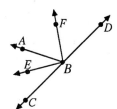

11. The bisectors of adjacent supplementary angles are perpendicular.

Given: $\angle ABC$ and $\angle ABD$ are supplementary,
$\overrightarrow{BE}$ bisects $\angle ABC$,
$\overrightarrow{BF}$ bisects $\angle ABD$

Prove: $\overrightarrow{BE} \perp \overrightarrow{BF}$

Proof:

Statements	Reasons
1. ?	1. Given
2. $m\angle EBC = m\angle ABE$, $m\angle ABF = m\angle FBD$	2. ? ∡ Ang. Bis ∠ Def
3. $m\angle ABC = m\angle ABE +$ $m\angle EBC$ $m\angle ABD = m\angle ABF +$ $m\angle FBD$	3. ? ∡ Add Def
4. $m\angle ABC = 2m\angle ABE$ $m\angle ABD = 2m\angle ABF$	4. ? Sub.
5. $m\angle ABC + m\angle ABD =$ $2m\angle ABE + 2m\angle ABF$	5. ?
6. $m\angle ABC + m\angle ABD = 180°$	6. ?
7. $2m\angle ABE + 2m\angle ABF = 180°$	7. ?
8. $m\angle ABE + m\angle ABF = 90°$	8. ?
9. $\overrightarrow{BE} \perp \overrightarrow{BF}$	9. ?

Figure for Exercise 12

12. The sum of the measures of the supplements of complementary angles equals 270°.

Given: $\angle 3$ and $\angle 4$ are comp.,
$\angle 2$ and $\angle 3$ are supp.,
$\angle 1$ and $\angle 4$ are supp.

Prove: $m\angle 1 + m\angle 2 = 270°$

Proof:

Statements	Reasons
1. ?	1. Given
2. ?	2. Def. of comp.
3. ?	3. Def. of supp.
4. $m\angle 1 + m\angle 2 + m\angle 3$ $+ m\angle 4 = 360°$	4. ?

5. ?   5. Substitution

6. ?   6. Add. Axiom of =

Figure for Exercise 13

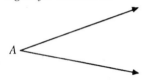

13. Every angle is congruent to itself. (Reflexivity)
   Given: ?
   Prove: ?
   Proof:

Statements	Reasons
1. ?	1. Given
2. $m\angle A = m\angle A$	2. ?
3. ?	3. Def. of $\simeq \angle$s

Figure for Exercise 14

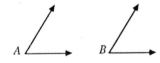

14. If $\angle A \simeq \angle B$, then $\angle B \simeq \angle A$. (Symmetry)
   Given: $\angle A \simeq \angle B$
   Prove: $\angle B \simeq \angle A$
   Proof:

Statements	Reasons
1. $\angle A \simeq \angle B$	1. ?
2. $m\angle A = m\angle B$	2. ?
3. $m\angle B = m\angle A$	3. ?
4. $\angle B \simeq \angle A$	4. ?

15. If $\angle A \simeq \angle B$ and $\angle B \simeq \angle C$, then $\angle A \simeq \angle C$. (Transitivity)

16. Complements of congruent angles are congruent.

2.5 CONGRUENT TRIANGLES

We now define a concept that allows us to compare triangles of the same *size* and *shape*. However, *size* and *shape* are geometrically illegal terms, since they have not been defined. Intuitively, what we mean by "same size and shape" is that if one triangle were placed on top of another triangle, the two triangles would coincide.

Definition

Two triangles, $\triangle ABC$ and $\triangle DEF$ are **congruent** if $\angle A \simeq \angle D$, $\angle B \simeq \angle E$, $\angle C \simeq \angle F$, $\overline{AB} \simeq \overline{DE}$, $\overline{BC} \simeq \overline{EF}$, and $\overline{AC} \simeq \overline{DF}$. We write $\triangle ABC \simeq \triangle DEF$.

Figure 2.8

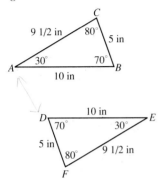

To find the congruence between two triangles, we must determine the one-to-one correspondence ($\longleftrightarrow$) between the vertices of the triangles so that they can be made to coincide. Thus, in Figure 2.8, $\triangle ABC \simeq \triangle EDF$ and $A \longleftrightarrow E$, (read A *corresponds to* E), $B \longleftrightarrow D$, and $C \longleftrightarrow F$. It would be incorrect to write $\triangle ABC \simeq \triangle DEF$, since $A \not\leftrightarrow D$ and $B \not\leftrightarrow E$.

We also write a correspondence between the **parts** (angles and sides) of a triangle. Thus, if $\triangle ABC \simeq \triangle EDF$ we write $\overline{AB} \longleftrightarrow \overline{ED}$ since $\overline{AB} \simeq \overline{ED}$, and we write $\angle A \longleftrightarrow \angle E$ since $\angle A \simeq \angle E$. Angles and sides that correspond in this manner are **corresponding parts**. Thus, a restatement of the definition of congruent triangles is:

Corresponding parts of congruent triangles are congruent.

This statement will often be used as a reason in proofs instead of the phrase "Def. of $\simeq$."

In order to avoid giving specific lengths of segments and measures of angles, as in Figure 2.8, we mark diagrams to indicate congruences, as in Figure 2.9, where the marks indicate that $\angle A \simeq \angle D$, $\angle B \simeq \angle E$, $\angle C \simeq \angle F$, $\overline{AB} \simeq \overline{DE}$, $\overline{AC} \simeq \overline{DF}$, and $\overline{BC} \simeq \overline{EF}$.

Figure 2.9

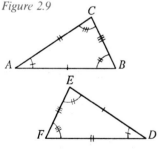

In a triangle, an **angle is included between two sides** if the sides are contained in the angle. **A side is included between two angles** if the side is contained in both angles. Thus, in Figure 2.10, given $\overline{AC}$ and $\overline{AB}$, $\angle A$ is the included angle, since $\overline{AC} \cup \overline{AB} \subset \angle A$, and, given $\angle A$ and $\angle B$, $\overline{AB}$ is the included side, since $\overline{AB} \subset \angle A \cap \angle B$.

Although the definition of congruent triangles states that all six parts of one triangle must be congruent to the corresponding parts of the other triangle, the next three postulates allow us to prove congruence given less information. If we were interested in a rigorous treatment of plane geometry, we could prove Postulates 2.2 and 2.3, making them theorems, but for simplicity, we choose to assume their validity.

Figure 2.10

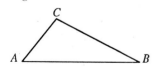

Postulate
2.1

(SAS = SAS)   If two sides and the included angle of one triangle are congruent to the corresponding parts of another triangle, then the triangles are congruent.

Postulate 2.2	(ASA = ASA) If two angles and the included side of one triangle are congruent to the corresponding parts of another triangle, then the triangles are congruent.

Postulate 2.3	(SSS = SSS) If three sides of one triangle are congruent to the corresponding parts of another triangle, then the triangles are congruent.

Figure 2.11

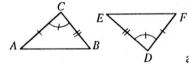

In Figure 2.11, $\triangle ABC \simeq \triangle FED$ by SAS = SAS.

The following example illustrates the use of Postulate 2.2 within a proof.

Example 1

Given: $\angle A \simeq \angle E, \overline{AC} \simeq \overline{CE}, A-C-E, B-C-D$

Prove: $\angle B \simeq \angle D$

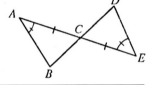

Solution

Proof:

Statements	Reasons
1. $\angle A \simeq \angle E, \overline{AC} \simeq \overline{CE}$	1. Given
2. $\angle ACB \simeq \angle ECD$	2. Vert. $\angle$ s are $\simeq$
3. $\triangle ACB \simeq \triangle ECD$	3. ASA = ASA
4. $\angle B \simeq \angle D$	4. Corr. parts of $\simeq \triangle$ s are $\simeq$

The following example illustrates the use of Postulate 2.3 within a proof.

Example 2

Given: $AB = CD, AD = CB$

Prove: $\angle A \simeq \angle C$

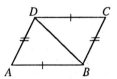

Solution

Proof:

Statements	Reasons
1. $AB = CD$, $AD = CB$	1. Given
2. $BD = DB$	2. Reflexivity
3. $\triangle ABD \simeq \triangle CDB$	3. SSS = SSS
4. $\angle A \simeq \angle C$	4. Corr. parts of $\simeq \triangle$s are $\simeq$

EXERCISE SET 2.5

In Exercises 1–3, if $\triangle ACB \simeq \triangle FED$, find the correspondences between:

1. The vertices of the triangles

2. The sides of the triangles

3. The angles of the triangles

For each pair of triangles in Exercises 4–5, write the congruence between the triangles.

4.

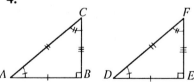

5.

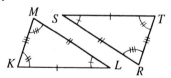

In Exercises 6–9, let $\triangle IJK \simeq \triangle PQR$.

6. Why is $\overline{IJ} \simeq \overline{PQ}$?

8. Why is $JK = QR$?

7. Why is $\angle K \simeq \angle R$?

9. Why is $m\angle KIJ = m\angle RPQ$?

For each of the Exercises 10–19, determine whether or not there is a congruence between $\triangle ABC$ and $\triangle DEF$. If there is such a congruence, write it and list the reasons for your answers (SAS = SAS, ASA = ASA, or SSS = SSS). If no such congruence exists, write *None*.

10. $\angle A \simeq \angle D$, $AC = DF$, $AB = DE$

11. $\angle A \simeq \angle E$, $BC = DF$, $\angle C \simeq \angle D$

12. $AC = DF$, $BC = FE$, $DE = AB$

13. $m\angle A = m\angle F$, $m\angle B = m\angle E$, $m\angle C = m\angle D$

14. $m\angle C = m\angle F$, $\angle B \simeq \angle E$, $\overline{BC} \simeq \overline{EF}$

15.

 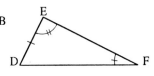

16.

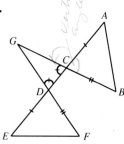

17.

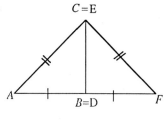

18.

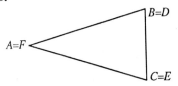

19.

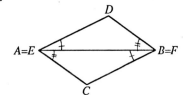

20. *Given:* $\angle A \simeq \angle Q$,
    $AC = QR$,
    $AB = PQ$
    Prove: $BC = PR$

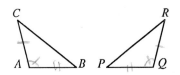

21. *Given:* $AD = BC$,
    $AB = CD$
    Prove: $\angle A \simeq \angle C$

22. *Given:* $AD = BC$,
    $\angle ADB \simeq \angle CBD$
    Prove: $\triangle ADB \simeq \triangle CBD$

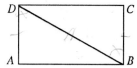

23. *Given:* $AC = AD$,
$\overline{AB} \perp \overline{CD}$
$\angle 1 \simeq \angle 2$

    Prove: $\triangle ABC \simeq \triangle ABD$

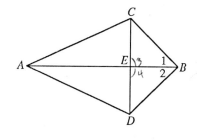

24. *Given:* $\overline{AD} \perp \overline{BF}$, $A-B-D$
$\angle 1 \simeq \angle 2$,
$AB = BD$,
$\angle 3 \simeq \angle 4$

    Prove: $\triangle ABC \simeq \triangle DBE$ ASA

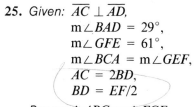

25. *Given:* $\overline{AC} \perp \overline{AD}$,
$m\angle BAD = 29°$,
$m\angle GFE = 61°$,
$m\angle BCA = m\angle GEF$,
$AC = 2BD$,
$BD = EF/2$

    Prove: $\triangle ABC \simeq \triangle FGE$

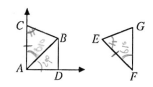

26. *Given:* $\angle 1$ and $\angle 2$ are supp.,
$\angle 3 \simeq \angle 4$,
$HG = EF$,
$DG = BF$

    Prove: $DH = BE$

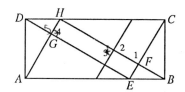

27. A carpenter who wishes to support a vertical wall will often nail a brace from the top of the wall to a position on the floor away from the wall. Which postulate states that the wall will now stay in place?

28. Explain how Postulates 2.1–2.3 relate to the fact that many structures, such as buildings, make use of triangular supports.

2.6 ISOSCELES TRIANGLES

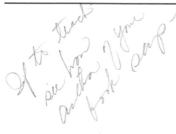

In the proofs of Theorems 2.1 and 2.3, we alternated between statements asserting that two angles are congruent and statements asserting that the measures of two angles are equal. In the future, for simplicity, we will treat the statement "$\angle 2 \simeq \angle 4$" as identical to the statement "$m\angle 2 = m\angle 4$," without taking a step to give the reason. $\overline{AB} \simeq \overline{CD}$ will be taken as $AB = CD$, without a reason being stated.

Definition	A **scalene** triangle is a triangle in which no two sides are congruent.

In Figure 2.10, $\triangle ABC$ is scalene.

Definition	An **isosceles** triangle is a triangle with at least two congruent sides. The angle included between two congruent sides is a **vertex angle**. The other two angles are **base angles** and the side opposite the vertex angle is the **base**.

Definition	A triangle with all sides congruent is **equilateral**. A triangle with all angles congruent is **equiangular**.

Figure 2.12

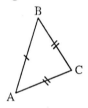

In Figure 2.12, $\triangle ABC$ is isosceles with a vertex angle $\angle C$, base angles $\angle A$ and $\angle B$, and base $\overline{AB}$. Also, $\triangle DEF$ is equilateral and $\triangle GHI$ is equiangular.

Theorem 2.4	If two sides of a triangle are congruent, the angles opposite these sides are congruent.

Given: $\triangle ABC, \overline{AC} \simeq \overline{BC}$

Prove: $\angle A \simeq \angle B$

Proof:

Statements	Reasons
1. $\overline{AC} \simeq \overline{BC}$	1. Given
2. $\overline{BC} \simeq \overline{AC}$	2. Symmetry
3. $\angle C \simeq \angle C$	3. Reflexivity
4. $\triangle ABC \simeq \triangle BAC$	4. SAS = SAS
5. $\angle A \simeq \angle B$	5. Corr. parts of $\simeq \triangle$s are $\simeq$

The above proof may seem a little strange, since it seems to involve only one triangle; however, keep in mind the possibility of needing to compare a given triangle with itself using several different correspondences. An alternative to this procedure is to make a second drawing of the triangle, a technique illustrated in Example 2 following.

A **corollary** is a theorem that is closely related to another theorem. The following corollary follows easily from Theorem 2.4. The proof is left as an exercise.

Corollary

If a triangle is equilateral, it is equiangular.

The following theorem is the converse of Theorem 2.4 and is proved in a similar manner. The proof is left as an exercise.

Theorem 2.5

If two angles of a triangle are congruent, the sides opposite the angles are congruent.

The following example illustrates the use of Theorem 2.4 within a proof.

Example 1

Given: *A, B, D, E* are on the same line
 CD = CE, AD = BE

Prove: △*ABC* is isosceles

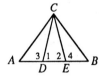

Solution

Analysis: To prove a triangle is isosceles, we need to show that two sides are congruent. One way to do this is to show that the sides are corresponding parts of congruent triangles. In order to use a congruence theorem, we need the appropriate parts in △*ADC* and △*BEC*. Since these are not given directly, we can relate ∠3 and ∠4 to ∠1 and ∠2 by using supplementary angles. Using the appropriate theorem, notice that ∠1 ≃ ∠2, since *CD = CE* is given.

Proof:

Statements	Reasons
1. $CD = CE$	1. Given
2. $\angle 1 \simeq \angle 2$	2. If 2 sides of a $\triangle$ are $\simeq$, the $\angle$s opp. these sides are $\simeq$
3. A, B, D, E are on the same line	3. Given
4. $\angle 1$ and $\angle 3$ are supp. $\angle 2$ and $\angle 4$ are supp.	4. Def. of supp. and measure
5. $\angle 3 \simeq \angle 4$	5. Supps. of $\simeq \angle$s are $\simeq$
6. $AD = BE$	6. Given
7. $\triangle ADC \simeq \triangle BEC$	7. Statements 1, 5, and 6; SAS = SAS
8. $AC = BC$	8. Corr. parts of $\simeq \triangle$s are $\simeq$
9. $\triangle ABC$ is isosceles	9. Def. of isosceles

■

Sometimes a theorem may be illustrated by a diagram in which the triangles are **overlapping**. To simplify the analysis of the proof of the theorem, draw a separate diagram in which the triangles are not overlapping. The following example illustrates this concept.

■

Example 2

Given: A, D, C, F are on the same line
$\angle A \simeq \angle F$, $\angle 1 \simeq \angle 2$, $AD = CF$

Prove: $\triangle ABC \simeq \triangle FED$

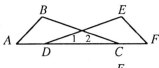

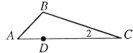

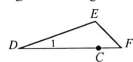

Solution

Proof:

Statements	Reasons
1. $\angle A \simeq \angle F$, $\angle 1 \simeq \angle 2$ $AD = CF$	1. Given
2. $DC = DC$	2. Reflexivity
3. $AD + DC = DC + CF$	3. Add. Axiom of $=$
4. $AD + DC = AC$ $DC + CF = DF$	4. Def. of between
5. $AC = DF$	5. Substitution
6. $\triangle ABC \simeq \triangle FED$	6. ASA = ASA

■

EXERCISE SET 2.6

1. Is an isosceles triangle always equilateral? Why?

2. Is an equilateral triangle always isosceles? Why?

Find x and y for each of Exercises 3–8.

3.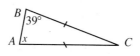

4.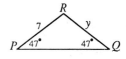

5.

6.

7.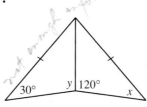

8.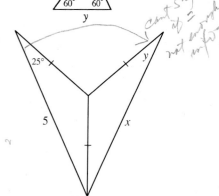

9. *Given:* $AB = BC$
   Prove: $\angle 1 \cong \angle 4$.

10. *Given:* $m\angle 1 + m\angle 3 = 180°$
    Prove: $AB = BC$.

Figure for Exercises 9–10

Figure for Exercises 11–12

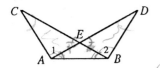

11. *Given:* $\triangle ABC \cong \triangle EDC$,
    $\angle 1 \cong \angle 2$
    Prove: $\triangle ABC$ is isosceles.

12. *Given:* $AC = CE, A-C-E, B-C-D,$
    $\triangle ABC$ and $\triangle DCE$ are equilateral
    Prove: $\triangle ABD \cong \triangle DEA$

13. *Given:* $PQ = PR,$
    $\angle 3 \cong \angle 4$
    Prove: $\triangle QPS \cong \triangle RPT$

14. *Given:* $\angle 1 \cong \angle 2, Q-S-T-R,$
    $QS = TR$
    Prove: $\triangle QPT \cong \triangle RPS$

Figure for Exercises 13–14

Figure for Exercises 15–16

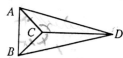

15. *Given:* $AE = BE,$
    $AC = BD,$
    $\angle 1 \cong \angle 2$
    Prove: $CE = DE$

16. *Given:* $\triangle ABC \cong \triangle BAD$
    Prove: $\triangle ABE$ is isosceles

17. *Given:* $AC = BC,$
    $AD = BD$
    Prove: $\angle DAC \cong \angle DBC$

18. *Given:* $AC = BC,$
    $\angle ACD \cong \angle BCD$
    Prove: $\triangle ABD \cong \triangle BAD$

Figure for Exercises 17–18

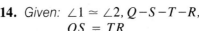

19. Prove that if a triangle is equilateral, it is equiangular.

20. Prove Theorem 2.5.

21. Prove that if a triangle is equiangular, it is equilateral.

2.7 ALTITUDES AND MEDIANS

In some proofs, segments that are not shown as parts of the original diagram must be added to the diagram. These are usually drawn with dashes and are called **auxiliary** segments. In the following example, the dashes are part of the solution, not part of the original problem.

Example 1

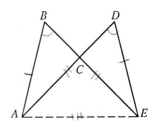

Given: $\overline{AB} \simeq \overline{ED}$, $\overline{BE} \simeq \overline{DA}$

Prove: $\angle B \simeq \angle D$

Solution

Proof:

Statements	Reasons
1. $\overline{AB} \simeq \overline{ED}$, $\overline{BE} \simeq \overline{DA}$	1. Given
2. $\overline{AE} \simeq \overline{AE}$	2. Reflexivity
3. $\triangle ABE \simeq \triangle EDA$	3. SSS = SSS
4. $\angle B \simeq \angle D$	4. Corr. parts of $\simeq \triangle$s are $\simeq$

Figure 2.13

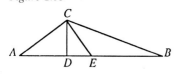

The following definitions will be easier to understand if you refer to Figure 2.13. In this figure it is not necessary to have $A - D - E - B$, but this arrangement is one of the possibilities.

Definition

Let $\triangle ABC$ be given with D and E on $\overleftrightarrow{AB}$. Then $\overline{CD}$ is an **altitude** of $\triangle ABC$ if $\overline{CD} \perp \overleftrightarrow{AB}$, and $\overline{CE}$ is a **median** of $\triangle ABC$ if $AE = BE$.

As described above, an altitude of a triangle need not intersect the opposite side. The only requirement is that the altitude be perpendicular to the line containing the opposite side. In Figure 2.14, $\overline{CD}$, $\overline{AE}$, and $\overline{BF}$ are altitudes of $\triangle ABC$.

Definition

> A **right triangle** is a triangle with an angle that is a right angle. The side opposite the right angle is the **hypotenuse** and the other two sides are **legs**.

Figure 2.14

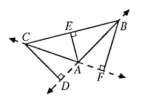

Thus, in Figure 2.14, $\triangle ABE$ is a right triangle, $\overline{AB}$ is the hypotenuse, and $\overline{AE}$ and $\overline{BE}$ are legs.

Referring again to Figure 2.14, it appears that every triangle has three altitudes. This is indeed true; however, this observation depends on the statement "there is exactly one perpendicular to a line from a point not on the line," that is, $\overline{AE}$ is the only altitude of $\triangle ABC$ containing the point A. To prove this, we first need another theorem.

Theorem 2.6

> An exterior angle of a triangle is greater in measure than either remote interior angle.

Given: $\triangle ABC$, $A-B-D$

Prove: (a) $m\angle CBD > m\angle C$
(b) $m\angle CBD > m\angle A$

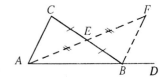

Note: In this proof, we will need auxiliary points and segments. The existence of such points and segments must be justified in the proof.

Proof (a):

Statements	Reasons
1. Let E be the midpoint of $\overline{BC}$	1. Ruler Post. and def. of segment
2. $CE = BE$	2. Def. of midpoint
3. Let F be a point such that $A-E-F$ and $AE = EF$	3. Ruler Post. and def. of between
4. $\angle CEA \simeq \angle BEF$	4. Vert. $\angle$s are $\simeq$
5. $\triangle ACE \simeq \triangle FBE$	5. SAS = SAS
6. $m\angle C = m\angle FBE$	6. Corr. parts of $\simeq \triangle$s are $\simeq$

7. $m\angle CBD = m\angle FBE$ $+ m\angle FBD$	7. Angle Add. Post.
8. $m\angle CBD = m\angle C + m\angle FBD$	8. Substitution
9. $m\angle CBD - m\angle C = m\angle FBD$	9. Add. Axiom of =
10. $m\angle FBD > 0$	10. Protractor Post.
11. $m\angle CBD - m\angle C > 0$	11. Substitution
12. $m\angle CBD > m\angle C$	12. Add. Axiom of Inequality

Proof (b): (Exercise Set 2.7, No. 18).

It is often convenient to rephrase a theorem in order to facilitate its proof. The following postulate is very useful in this regard.

Postulate
2.4

> A statement and its contrapositive are either both true or both false.

Example 2

Discuss the truth or falsity of the contrapositive of the statement "If it is cloudy, then it is raining."

Solution

The statement given is false, since sometimes when it is cloudy, it is not raining. Therefore, by Postulate 2.4, the contrapostive of the statement is false. ("If it is not raining, then it is not cloudy.")

■

By Postulate 2.4, in order to prove a statement, one can instead prove its contrapositive. We shall develop this concept more fully in Section 3.2, through the process of *indirect proof*. The following theorem is an example of a statement that is proved by proving the contrapositive is true.

Theorem
2.7

> There is exactly one perpendicular to a line that contains a point not on the line.

Figure 2.15

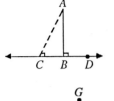

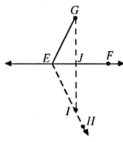

This theorem can be proved by examining the concepts of **existence** and **uniqueness**. To prove that there is *exactly one* perpendicular satisfying a condition, we may prove that there is *at least one* such perpendicular (existence) and that there is *at most one* such perpendicular (uniqueness).

The detailed proof of Theorem 2.7 is left as an exercise (Exercise Set 2.7, No. 19); however, we give an analysis of the proof here. The uniqueness portion of the theorem is proved by showing that if Theorem 2.7 is not true, then Theorem 2.6 is not true. See $\overleftrightarrow{AB}$ and $\overleftrightarrow{CD}$ in Figure 2.15. The existence portion of the theorem is proved by constructing an angle $\angle FEH$ congruent to $\angle FEG$ such that G and H are on opposite sides of $\overleftrightarrow{EF}$ and then finding a point I on $\overrightarrow{EH}$ such that $EI = EG$. Line segment $\overline{GI}$ must intersect line $\overleftrightarrow{EF}$ at point J. Using congruence theorem methods, it can be shown that $\triangle EGJ \simeq \triangle EIJ$ and that $\angle EJG \simeq \angle EJI$ and, thus, $\overline{GI} \perp \overleftrightarrow{EF}$. See Figure 2.15.

EXERCISE SET 2.7

1. What is the difference between the altitude of a triangle and the median of a triangle?

2. How many altitudes does a triangle have? How many medians does a triangle have?

3. Construct a scalene triangle and construct all its altitudes and medians.

4. Construct an isosceles triangle, and construct all its altitudes and all its medians.

5. Construct a right triangle, and construct all its altitudes and all its medians.

6. Do all the medians of a triangle seem to intersect in a single point?

7. If lines containing the altitudes of a triangle are drawn, do they seem to intersect in a single point?

Write the contrapositive of each of the statements in Exercises 8–9.

8. If two angles of a triangle are not congruent, then the sides opposite these angles are not congruent.

9. If $m\angle 1 \neq m\angle 2$, then $AB \neq CD$.

10. *Given:* $AC = BC$,
    $\overline{CD}$ is a median

    Prove: $\overline{CD}$ is an altitude

11. *Given:* $\overline{CD}$ is an altitude,
    $\overline{CD}$ is a median

    Prove: $\triangle ACD \simeq \triangle BCD$

Figure for Exercises 10-11

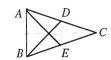

Figure for Exercises 12-13

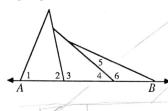

12. *Given:* line $\overleftrightarrow{AB}$

    Prove: $m\angle 2 > m\angle 5$

13. *Given:* line $\overleftrightarrow{AB}$

    Prove: $m\angle 1 < m\angle 6$

14. *Given:* $AC = BC$,
    $\overline{AE}$ and $\overline{BD}$ are medians of $\triangle ABC$

    Prove: $AE = BD$

Figure for Exercises 14-15

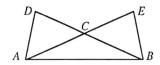

15. *Given:* $AC = BC$,
    $\angle DAE \simeq \angle EBD$,
    $AD = CE$

    Prove: $\overline{AE}$ and $\overline{BD}$ are medians of $\triangle ABC$

16. *Given:* $m\angle E = 90°$,
    $CD = EC$,
    $m\angle CAB = m\angle CBA$

    Prove: $\triangle ABD$ is a right triangle

Figure for Exercises 16-17

17. *Given:* $AD = BE$,
    $CD = CE$,
    $\overline{AD}$ and $\overline{BE}$ are altitudes of $\triangle ABC$

    Prove: $\triangle ABC$ is isosceles

18. *Prove:* Theorem 2.6 (*b*).

19. *Prove:* Theorem 2.7.

20. Listen to conversations around you. When you hear an "if-then" statement, write it down. Write the contrapositive of the statement. Does it have the same meaning?

21. Listen to conversations around you. When you hear an "if-then" statement, write it down. Write the converse of the statement. Does it have the same meaning?

Chapter 2
Key Terms

Chapter 2
Review Exercises

Fill each blank with the word *always, sometimes,* or *never.*

1. The bisectors of a pair of adjacent supplementary angles are _____ perpendicular to each other.

2. Supplements of two angles are _____ congruent.

3. The bisectors of supplementary angles are _____ perpendicular.

4. If three angles of one triangle are congruent to the corresponding parts of another triangle, then the triangles are _____ congruent.

5. An altitude of a triangle is _____ a median of the triangle.

6. If a statement is true, its contrapositive is _____ false.

7. If a statement is true, its converse is _____ false.

8. Two right triangles are _____ congruent.

9. An equilateral triangle is _____ isosceles.

10. A median of a triangle _____ divides it into two congruent triangles.

True-False: If the statement is true, mark it so. If it is false, replace the underlined word to make a true statement.

11. The hypothesis of a theorem <u>includes</u> the word "if."

12. Supplements of <u>adjacent</u> angles are congruent.

13. Vertical angles are <u>complementary.</u>

14. Corresponding parts of congruent triangles are <u>congruent.</u>

15. If three <u>angles</u> of one triangle are congruent to the corresponding parts of another triangle, the triangles are congruent.

16. <u>A scalene</u> triangle is a triangle with at least two congruent sides.

17. <u>A corollary</u> is a theorem that is closely related to another theorem.

18. <u>An altitude</u> of a triangle always bisects one side of the triangle.

19. An exterior angle of a triangle is <u>smaller</u> in measure than either remote interior angle.

20. There is exactly one <u>median</u> to a line that contains a point not on the line.

Answer the following questions.

21. Write the hypothesis: "It is cloudy if it is raining."

22. Write the conclusion: "Buckle up for safety."

23. Write the converse: "Isosceles triangles are not scalene."

24. Write the inverse: "If $AB = BC$, then $m\angle A \neq m\angle B$."

25. Write the contrapositive: "Beets are a treat when the sugar is sweet."

26. Name all pairs of vertical angles in the figure.

27. Name all pairs of supplementary angles in the figure.

28. Name all pairs of congruent angles in the figure.

Figure for Exercises 26–28

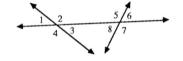

29. In $\triangle ABC$, if $\angle A \simeq \angle B$ and $AC = 10$, find BC.

30. If $\overrightarrow{AD}$ bisects $\angle BAC$ and $m\angle BAC = 37°$, find $m\angle BAD$.

31. If $\overleftrightarrow{AB} \cap \overleftrightarrow{CD} = \{E\}$ and $m\angle AEC = 112°$, find $m\angle AED$.

32. If $\triangle QRS \simeq \triangle VRW$, find the correspondence between the vertices of the triangles, the sides of the triangles, and the angles of the triangles.

In Exercises 33–36, state why $\triangle ABC \simeq \triangle DCB$.

33. $\angle A \simeq \angle D$, $m\angle ACB = m\angle DBC$, $AC = BD$.

34. $\angle ACB \simeq \angle DBC$, $\angle ACD \simeq \angle ABD$

35. $AC = BD$, $AB = CD$

36. $AB = CD$, $m\angle ABC = m\angle DCB$

37. If $\triangle ABC$ is equiangular and $AC = 23$, find AB and BC.

38. If $\triangle ABC$ is equilateral and $AC = 23$, find AB and BC.

39. In $\triangle ABC$, if $A-D-B$ and $m\angle BDC = 72°$, what can be said about $m\angle ACD$?

40. If the statement "All apples are kangaroos" is true, explain why the following statement is true: "If something is not a kangaroo, then it is not an apple."

Proofs:

In Exercises 41–46, let $A-D-E-B$, $A-F-C$, and $B-G-C$.

41. *Given:* $AD = BE$
    Prove: $AE = BD$

Figure for Exercises 41–46

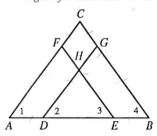

In Exercises 42–43, given: $AF = EF = DG = BG, AD = BE, CF = CG$

42. *Prove:* $\angle 1 \simeq \angle 2 \simeq \angle 3 \simeq \angle 4$
43. *Prove:* $\triangle ABC$ and $\triangle DEH$ are isosceles

In Exercises 44–46, given: $AF = EF = DG = BG, AD = BE$

44. *Prove:* $\triangle AFE \simeq \triangle DGB$ by using SSS = SSS

45. *Prove:* $\triangle AFE \simeq \triangle BGD$ by using ASA = ASA

46. *Prove:* $\triangle FEA \simeq \triangle GDB$ by using SAS = SAS

47. *Given:* $B-D-C$,
      $AB = AC$,
      $BD = CD$
  Prove: $\overrightarrow{AD}$ bisects $\angle BAC$

Figure for Exercises 47–48

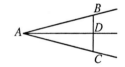

48. *Given:* $B-D-C$,
      $AB = AC$,
      $\overrightarrow{AD}$ bisects $\angle BAC$
  Prove: $\overline{BC} \perp \overline{AD}$

49. *Given:* $A-D-B$,
      $C-E-D$,
      $AC = BC$,
      $AE = BE$
  Prove: $\overline{CD} \perp \overline{AB}$ and $\overline{CD}$ is a median

Figure for Exercise 49–50

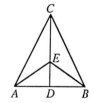

50. *Given:* $A-D-B$,
      $C-E-D$,
      $\overline{CD} \perp \overline{AB}$,
      $AD = BD$
  Prove: $\triangle ABC$ and $\triangle ABE$ are isosceles

Figure for Exercises 51–52

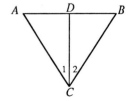

51. *Given:* $A-D-B$,
      $m\angle 1 = m\angle 2$,
      $\overline{CD}$ is an altitude
  Prove: $\overline{CD}$ is a median

52. *Given:* $A-D-B$,
$\qquad\quad$ $AC = BC$
$\qquad\quad$ $\overline{CD}$ is a median,
$\quad$ *Prove:* $\overline{CD}$ is an altitude

Constructions:

53. Draw a fairly large scalene triangle with acute angles, and construct its medians. (These should all intersect in one point.)

54. Draw a fairly large scalene triangle with acute angles, and construct its altitudes.

55. Repeat Exercise 53 with one angle obtuse.

56. Repeat Exercise 54 with one angle obtuse.

start and med Tim

Parallels and Polygons

Courtesy of Historical Pictures Service, Chicago

Pythagoras (?–ca. 497 B.C.)

Historical Note

The Greek mathematician Pythagoras gathered a group of students together in Crotona, Italy, to do research work in geometry, astromony, and music. The Pythagorean school used the terms *figurate* and *polygonal* to represent certain sets of numbers that could be illustrated by geometric patterns. For example, 1, 3, and 6 are triangular numbers

```
                              *
              *           *   *
  *       *   *       *   *   *
```

and 1, 4, and 9 are square numbers.

```
                      *   *   *
          *   *       *   *   *
  *       *   *       *   *   *
```

Other common sets are pentagonal and hexagonal numbers.

In this chapter, we present the concept of parallel lines in Euclidean geometry. We then extend the methods of proof to include indirect proofs. Various types of polygons and theorems relating to polygons are discussed, including polygons containing parallel sides.

3.1 PARALLEL POSTULATES

Two lines in the same plane will either intersect or fail to intersect. We have already discussed intersecting lines; we now examine nonintersecting lines.

Definition | Two distinct lines in a plane are **parallel** if they do not intersect.

Two distinct line segments are parallel if the lines containing them are parallel. If line j is parallel to line k, we write $j \parallel k$, and say "j is parallel to k." Using this notation we can say that $j \parallel k$ if $j \cap k = \emptyset$.

Figure 3.1

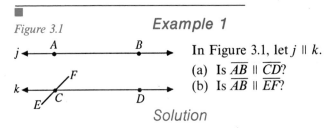

Example 1

In Figure 3.1, let $j \parallel k$.
(a) Is $\overline{AB} \parallel \overline{CD}$?
(b) Is $\overline{AB} \parallel \overline{EF}$?

Solution

(a) Notice that $\overline{AB} \subset j$ and $\overline{CD} \subset k$, so we know that $\overline{AB} \parallel \overline{CD}$.
(b) $\overline{AB}$ is not parallel to $\overline{EF}$, even though $\overline{AB} \cap \overline{EF} = \emptyset$, since the lines containing $\overline{AB}$ and $\overline{EF}$ are not parallel.

A point is **outside** a set if it is not an element of the set.

Example 2

In Figure 3.2,
(a) Is A outside line m?
(b) Is P outside $\triangle BCD$?

Figure 3.2

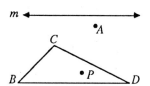

Solution

(a) Yes, since *A* is not an element of line *m*.
(b) Yes, since *P* is not an element of △*BCD*.

■

A line passes **through** a point if the line contains the point.

Postulate
3.1

> **(Parallel Postulate)**   Through a point outside a given line there is exactly one parallel to the line.

■

Example 3

Figure 3.3

In Figure 3.3, if *p* ∥ *n*, how many lines are there through *Q* parallel to *n*?

Solution

By Postulate 3.1, *p* is the only line through *Q* parallel to line *n*.

■

The Parallel Postulate caused a great deal of controversy for many centuries. Euclid was unable to prove the statement and, therefore, made it a postulate. Many mathematicians tried to prove it, but failed. Finally, several mathematicians, working independently but at nearly the same time, decided to replace this postulate with a different one, and in so doing developed new (non-Euclidean) geometries. Lobachevsky and Bolyai developed the *L-B Postulate*, and Riemann, developed the *R Postulate*.

L-B Postulate

> Through a point outside a given line there are infinitely many parallels to the line.

R Postulate

> Through a point outside a given line there are no parallels to the line.

Neither the L-B Postulate nor the R Postulate will be used by us at this time. For further discussion of non-Euclidean geometries, see Chapter 10.

EXERCISE SET 3.1

Answer the questions and give reasons for your answers. In Exercises 1–5, $\overleftrightarrow{AB} \parallel \overleftrightarrow{CD}$ in the given figure.

Figure for Exercises 1–5

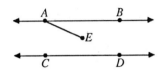

1. Is $\overline{AB} \parallel \overline{CD}$?

2. Is $\overline{AE} \parallel \overline{AB}$?

3. Is $\overline{AE} \parallel \overline{CD}$?

4. Does there exist a point P on $\overleftrightarrow{CD}$ such that $\overleftrightarrow{EP} \parallel \overleftrightarrow{BD}$?

5. Does there exist a point Q on $\overleftrightarrow{CD}$ such that $\overleftrightarrow{EQ} \parallel \overleftrightarrow{AB}$?

In Exercises 6–10, let j, k, and m be three distinct lines and let P be a point.

6. If $j \parallel k$ and $j \parallel m$, is $k \parallel m$?

7. If $j \parallel k$ and $k \cap m \neq \emptyset$, is $j \parallel m$?

8. If $j \cap k \neq \emptyset$ and $k \cap m \neq \emptyset$, is it possible that $j \cap m = \emptyset$?

9. If $j \cap k = \{P\}$ and $k \cap m \neq \{P\}$, is $j \cap m = \{P\}$?

10. If $j \cap k \neq \emptyset$, $j \cap k \neq \{P\}$, $k \cap m = \{P\}$, is it possible that $j \cap m = \{P\}$?

11. Give examples of parallel line segments on a football field.

12. Give examples of parallel line segments on a tennis court.

13. Give examples of line segments that are not parallel on a baseball field.

14. Give examples of line segments that are not parallel on a basketball court.

3.2 INDIRECT PROOF

By the **law of excluded middle**, if two statements are contradictory and one is true, then the other is false. The indirect method of proof uses this law to prove certain theorems.

Consider the two statements $\angle 1 \simeq \angle 2$ and $\angle 1 \not\simeq \angle 2$. These statements are contradictory, since $\angle 1 \simeq \angle 2$ and $\angle 1 \not\simeq \angle 2$ cannot both be true. If $\angle 1 \simeq \angle 2$ is true, then $\angle 1 \not\simeq \angle 2$ must be false, and conversely.

In a given problem, suppose we wish to prove that $AB \neq CD$. We can do this by proving that the statement $AB = CD$ is false. We do so by first assuming that the statement $AB = CD$ is true. If this assumption leads to a contradiction of fact, then the conclusion that $AB = CD$ must be false, and the conclusion $AB \neq CD$ is true.

We will prove the next theorem using **indirect proof.**

Theorem 3.1

If two angles of a triangle are not congruent, the sides opposite these angles are not congruent.

Given: $\triangle ABC$, $\angle C \neq \angle A$

Prove: $\overline{AB} \not\simeq \overline{BC}$

Proof:

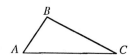

Statements	Reasons
1. Either $\overline{AB} \simeq \overline{BC}$ or $\overline{AB} \not\simeq \overline{BC}$	1. Law of excluded middle
2. $\overline{AB} \simeq \overline{BC}$	2. Assumption
3. $\angle C \simeq \angle A$	3. If 2 sides of a $\triangle$ are $\simeq$, the $\angle$s opp. these sides are $\simeq$
4. $\angle C \not\simeq \angle A$	4. Given
5. $\overline{AB} \not\simeq \overline{BC}$	5. Assuming $\overline{AB} \simeq \overline{BC}$ led to two contradictory statements (3 and 4); hence, the assumption (2) is false

The next theorem states more than the converse of Theorem 3.1 and can be proved without resorting to indirect proof.

Theorem 3.2

If two sides of a triangle are not congruent, the angles opposite these sides are not congruent, and the angle with larger measure is opposite the longer side.

Given: $\triangle ABC$, $BC > AB$

Prove: $m\angle A > m\angle C$

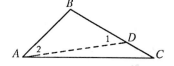

Proof:

Statements	Reasons
1. $BC > AB$	1. Given
2. There is a point D on $\overline{BC}$ such that $BD = AB$	2. Ruler Post. and def. of length
3. $m\angle 2 = m\angle 1$	3. If 2 sides of a $\triangle$ are $\simeq$, the $\angle$s opp. these sides are $\simeq$
4. $m\angle 1 > m\angle C$	4. An exterior $\angle$ of a $\triangle$ > either remote interior $\angle$
5. $m\angle A > m\angle 2$	5. Angle Add. Post. and def. of measure
6. $m\angle A > m\angle 1$	6. Substitution (State. 3, 5)
7. $m\angle A > m\angle C$	7. Transitivity (State. 4, 6)

Sometimes more than two mutually exclusive statements are involved in an indirect proof. In this case, to prove a statement true, we must prove that all other possibilities are false.

For example, we know that for any real numbers a and b, either $a < b$, $a = b$, or $a > b$. To prove that $a < b$, we must prove that the statements $a = b$ and $a > b$ are both false.

The converse of Theorem 3.2 can be proved in this manner. We shall prove it using **paragraph form** of a proof. Notice the similarity between this theorem and Theorem 3.1.

Theorem 3.3 | If two angles of a triangle are not congruent, the sides opposite these angles are not congruent, and the longer side is opposite the angle with larger measure.

Given: $\triangle ABC$, $m\angle A > m\angle C$

Prove: $BC > AB$

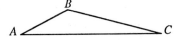

Proof: Either $BC = AB$, $BC < AB$, or $BC > AB$. Assume $BC = AB$. Then $m\angle A = m\angle C$, since if two sides of a triangle are congruent, the angles opposite these sides are congruent. The statement $m\angle A = m\angle C$ is a contradiction of the hypothesis $m\angle A > m\angle C$. Assume $BC < AB$. By Theorem 3.2, $m\angle A < m\angle C$. The statement $m\angle A < m\angle C$ is also a contradiction of the hypothesis $m\angle A > m\angle C$. Therefore, $BC > AB$.

The following is another example of a theorem that we shall prove using paragraph form.

Theorem 3.4 | Two lines in a plane perpendicular to a third line in the same plane are parallel.

Given: lines *i, j, k;*
           $i \perp j, i \perp k$

Prove: $j \parallel k$

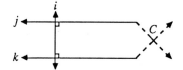

Proof: Either $j \parallel k$ or $j \cap k \neq \emptyset$. Assume that $j \cap k = \{C\}$. Then there are two perpendiculars to *i* through *C*, contradicting the statement that there is exactly one perpendicular to a line through a point not on the line (Theorem 2.7). Therefore, $j \parallel k$.

Example

Given: $\triangle ABC, \triangle DEC, AB = DE, BC = CE, AC \neq CD$

Prove: $m\angle B \neq m\angle E$

Solution

Either $m\angle B = m\angle E$ or $m\angle B \neq m\angle E$. Assume $m\angle B = m\angle E$. Then $\triangle ABC \simeq \triangle DEC$ by SAS = SAS, since we are given that $AB = DE$ and $BC = CE$. Thus, $AC = CD$, since corresponding parts of congruent triangles are congruent. This contradicts the given statement that $AC \neq CD$. Therefore, $m\angle B \neq m \angle E$.

EXERCISE SET 3.2

1. In $\triangle ABC$, if $m\angle A = 63°, m\angle B = 61°, m\angle C = 56°$, and $AC = 8$, what can be said about AB and BC? Why?

2. In $\triangle ABC$, if $m\angle A = 37°, m\angle B = 35°$, and $BC = 10$, what can be said about AC and AB? Why?

3. In $\triangle ABC$, if $AB = 5, BC = 7$, and $m\angle A = 42°$, what can be said about $m\angle C$? If $m\angle A + m\angle B + m\angle C = 180°$, what can be said about $m\angle B$? Why?

4. In $\triangle ABC$, if $AB = 10, BC = 12, AC = 14, m\angle A = 65°$, and if $m\angle A + m\angle B + m\angle C = 180°$, what can be said about $m\angle B$ and $m\angle C$? Why?

Figure for Exercises 5–6.

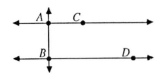

5. If $\overleftrightarrow{AC} \perp \overleftrightarrow{AB}$ and $\overleftrightarrow{BD} \perp \overleftrightarrow{AB}$, is $\overleftrightarrow{AC} \parallel \overleftrightarrow{BD}$? Is $\overline{AC} \parallel \overline{BD}$? Why?

6. If $\overleftrightarrow{AB} \perp \overleftrightarrow{AC}$ and $m\angle ACD \neq 90°$, is $\overleftrightarrow{CD} \parallel \overleftrightarrow{AB}$? Why?

7. Can a triangle have two parallel sides? Why?

8. Can a triangle have two right angles? Why?

Figure for Exercises 9–14

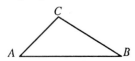

9. *Given:* $m\angle A \neq m\angle B$
Prove: $BC \neq AC$

10. *Given:* $AB = BC$, $m\angle A \neq m\angle B$
Prove: $AB \neq AC$

11. *Given:* $AB \neq BC$
Prove: $m\angle C \neq m\angle A$

12. *Given:* $m\angle B = m\angle A$, $AB \neq BC$
Prove: $m\angle B \neq m\angle C$

13. *Given:* $m\angle A \neq m\angle B$, $m\angle A \neq m\angle C$, $m\angle B \neq m\angle C$
Prove: $AB \neq BC$, $AB \neq AC$, $AC \neq BC$

14. *Given:* $AB \neq BC$, $AB \neq AC$, $AC \neq BC$
Prove: $m\angle A \neq m\angle B$, $m\angle A \neq m\angle C$, $m\angle B \neq m\angle C$
(A converse of Exercise 13)

Figure for Exercises 15–16

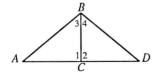

15. *Given:* $m\angle D < m\angle A < m\angle 1$
Prove: $BC < AB < BD$

16. *Given:* $AC = CD$,
$m\angle 3 \neq m\angle 4$
Prove: $m\angle 1 \neq m\angle 2$

Figure for Exercises 17–20

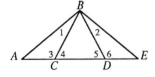

17. *Given:* $\triangle ABC \not\cong \triangle EBD$,
$AC = DE$
Prove: $m\angle 4 \neq m\angle 5$

18. *Given:* $AB = BE$,
$AC \neq DE$
Prove: $m\angle 1 \neq m\angle 2$

19. *Given:* $m\angle 4 \neq m\angle 5$,
$AC = DE$
Prove: $AB \neq BE$

20. *Given:* $m\angle 3 = m\angle 6$,
$m\angle 1 \neq m\angle 2$
Prove: $AC \neq DE$

3.3 TRANSVERSALS

A **transversal** is a line that intersects two or more lines in the same plane at distinct points. In Figure 3.4, line *t* is a transversal. Line *s* is not a transversal. Line *t* **cuts** lines *j* and *k* at points *A* and *B*, respectively.

Definition | The four angles formed by a pair of lines and a transversal, such that both points of intersection are on one of the sides of each angle are **interior angles**. The other four angles are **exterior angles**.

Figure 3.4(a)

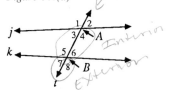

Figure 3.4(b)

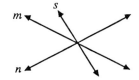

■
.

Example 1

In Figure 3.4(a) name the interior angles and the exterior angles.

Solution

Since angles 3, 4, 5, and 6 each contain both points *A* and *B*, they are interior angles. Angles 1, 2, 7, and 8 are exterior angles.

■

Definition | Two nonadjacent interior angles on opposite sides of a transversal are **alternate interior angles**. Two nonadjacent exterior angles on opposite sides of a transversal are **alternate exterior angles**.

■

Example 2

In Figure 3.4(a) name the alternate interior angles and the alternate exterior angles.

Solution

Angles 3 and 6 are alternate interior angles. Angles 4 and 5 are also alternate interior angles. Angles 1 and 8 are alternate exterior angles. Angles 2 and 7 are also alternate exterior angles.

■

Definition

Two nonadjacent angles on the same side of the transversal are **corresponding angles** if one angle is interior and one is exterior.

Example 3

In Figure 3.4(a) name the corresponding angles.

Solution

The corresponding angles are the pairs:
Angles 1 and 5, angles 3 and 7, angles 2 and 6, angles 4 and 8.

■

We state Theorem 3.5 for completeness, but leave the proof as an exercise.

Theorem 3.5

If a line in a plane is perpendicular to one of two parallel lines in the same plane, then it is also perpendicular to the other.

In most cases in this text, we use paragraph form to prove theorems that require an indirect proof. However, indirect proofs may be done in the manner illustrated in the proof of Theorem 3.1. All proofs can be done in paragraph form, but it is usually easier for the beginning student to learn how to do proofs in the standard two-column form. The following are two more examples of theorems proved by indirect proof using paragraph form.

Theorem
3.6

| If two lines are cut by a transversal so that alternate interior angles are congruent, then the lines are parallel. |

Given: m∠1 = m∠2

Prove: j ∥ k

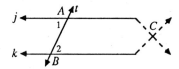

Proof: Either j ∥ k or j ∩ k ≠ 0. Assume that j ∩ k = {C}. Then ∠1 is an exterior angle of ΔABC and hence, m∠1 > m∠2, since the exterior angle of a triangle is greater in measure than either remote interior angle. The statement m∠1 > m∠2 contradicts the given statement m∠1 = m∠2; therefore, j ∥ k.

Theorem
3.7

| If two parallel lines are cut by a transversal, then the alternate interior angles are congruent. |

Given: $\overleftrightarrow{AB}$ ∥ $\overleftrightarrow{CD}$

Prove: m∠ABC = m∠DCB

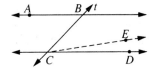

Proof: There are three possibilities: either m∠ABC < m∠DCB, m∠ABC > m∠DCB, or m∠ABC = m∠DCB. Assume m∠ABC < m∠DCB. Then there is a ray $\overrightarrow{CE}$ such that m∠ABC = m∠ECB. By Theorem 3.6, $\overleftrightarrow{AB}$ ∥ $\overleftrightarrow{CE}$. But by hypothesis $\overleftrightarrow{AB}$ ∥ $\overleftrightarrow{CD}$. Hence, through point C, there are two lines parallel to $\overleftrightarrow{AB}$, contradicting the Parallel Postulate. Likewise, the assumption m∠ABC > m∠DCB leads to a contradiction of the Parallel Postulate (see Exercise Set 3.3, No. 29); therefore, m∠ABC = m∠DCB.

Theorems 3.6 and 3.7 are converses of each other. We can state both theorems in one sentence using the words **if and only if**: Alternate interior angles formed by two lines and a transversal are congruent if and only if the lines are parallel. In general, if we are given two statements "p" and "q," the statement "p if and only if q" is equivalent to the statement "if p then q, and if q then p." The *if* portion of the statement is "if q then p" and the *only if* portion of the statement is "if p then q."

The following example is given to emphasize the difference between the "if statement" and the "only if statement." We intentionally give a simple example, so that the reader will concentrate on the format, and not necessarily the proof itself.

Example 4

Given: $\triangle ABC, \triangle DEF, AB = DE, BC = EF$

Prove: $\triangle ABC \simeq \triangle DEF$ if and only if $AC = DF$.

Solution

In (a) we prove the "if statement" and in (b) we prove the "only if statement."

(a) *Given:* $\triangle ABC, \triangle DEF, AB = DE, BC = EF, AC = DF$

   Prove: $\triangle ABC \simeq \triangle DEF$

   Proof:

Statements	Reasons
1. $\triangle ABC, \triangle DEF, AB = DE,$ $BC = EF, AC = DF$	1. Given
2. $\triangle ABC \simeq \triangle DEF$	2. SSS = SSS

(b) *Given:* $\triangle ABC \simeq \triangle DEF, AB = DE, BC = EF$

   Prove: $AC = DF$

   Proof:

Statements	Reasons
1. $\triangle ABC \simeq \triangle DEF$	1. Given
2. $AC = DF$	2. Corr. parts of $\simeq \triangle$s are $\simeq$

Notice that the given in part (b) contains more information than is necessary.

Theorem 3.8 | Corresponding angles formed by two lines and a transversal are congruent if and only if the lines are parallel.

(a) *Given:* $j \parallel k$

    Prove: $m\angle 1 = m\angle 3$

(b) *Given:* $m\angle 1 = m\angle 3$

    Prove: $j \parallel k$

Proof (a):

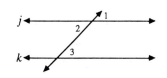

Statements	Reasons
1. $j \parallel k$	1. Given
2. $m\angle 1 = m\angle 2$	2. Vert. $\angle$s are $\simeq$
3. $m\angle 2 = m\angle 3$	3. If 2 $\parallel$ lines are cut by a trans., alt. int. $\angle$s are $\simeq$
4. $m\angle 1 = m\angle 3$	4. Transitivity

Proof (b): See Exercise Set 3.3, No 30.

The proofs of the following two theorems are left as exercises.

Theorem 3.9	Alternate exterior angles formed by two lines and a transversal are congruent if and only if the lines are parallel.

Theorem 3.10	Interior angles formed by two lines and a transversal, such that the angles are on the same side of the transversal, are supplementary if and only if the lines are parallel.

Example 5

In Figure 3.5, find the measures of the angles if m $\angle 3 = 55°$ and $j \parallel k$.

Figure 3.5

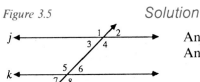

Solution

Angles 2, 3, 6, and 7 have measures of 55°.
Angles 1, 4, 5, and 8 have measures of 125°.

Figure 3.6

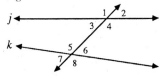

It is important for the reader to realize the necessity for the word *parallel* in the preceding theorems. In Figure 3.6, let $j \cap k \neq \emptyset$ and $m\angle 3 = 55°$. There is no way that we can find the measure of $\angle 6$, given only the measure of $\angle 3$. All we can be sure of is that $m\angle 6 \neq 55°$. Why?

Construction 3.1	Through a point outside a line construct the parallel to the line.

Given: $\overleftrightarrow{AB}$, *P*

Construct: $\overrightarrow{PR}$ ∥ $\overleftrightarrow{AB}$

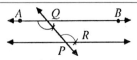

Step 1. Through *P*, draw any line intersecting $\overleftrightarrow{AB}$. Call the point of intersection *Q*.

Step 2. At *P*, construct ∠*QPR* such that ∠*QPR* ≃ ∠*PQA* and such that ∠*QPR* and ∠*PQA* are alternate interior angles.

Step 3. Draw $\overleftrightarrow{PR}$. This is the required line since, by the Parallel Postulate, there is exactly one parallel to $\overleftrightarrow{AB}$ through point *P*.

Other methods for this construction are possible. The following theorem is often used in computational problems.

Theorem 3.11	The sum of the measures of the angles of a triangle is 180°.

Given: △*ABC*

Prove: m∠1 + m∠*A* + m∠*C* = 180°

Proof:

Statements	Reasons
1. Construct $\overleftrightarrow{BD}$ ∥ $\overleftrightarrow{AC}$	1. Parallel Post.
2. ∠*EBD* is a straight ∠	2. Def. of straight ∠
3. m∠1 + m∠2 + m∠3 = m∠*EBD*	3. Angle Add. Post.
4. m∠1 + m∠2 + m∠3 = 180°	4. Def. of meas. of an ∠
5. m∠*A* = m∠2, m∠*C* = m∠3	5. If 2 ∥ lines are cut by a trans., alt. int ∠s are ≃.
6. m∠1 + m∠*A* + m∠*C* = 180°	6. Substitution

■

Example 6

If in $\triangle ABC$, $m\angle A = 57°$ and $m\angle C = 68°$, find $m\angle B$.

Solution

$m\angle B = 180° - 57° - 68° = 55°.$

■

■

Example 7

If in $\triangle ABC$, $m\angle A = 4(m\angle B)$ and $m\angle C = 5(m\angle B)$, find the measures of all the angles.

Solution

Let $m\angle B = x$. Then $m\angle A = 4x$ and $m\angle C = 5x$. Thus,

$$4x + x + 5x = 180°$$
$$10x = 180°$$
$$x = 18°.$$

Therefore, $m\angle A = 72°$, $m\angle B = 18°$, and $m\angle C = 90°$.

■

The proof of the final theorem in this section is left as an exercise.

Theorem 3.12	An exterior angle of a triangle is equal in measure to the sum of the remote interior angles.

■

Example 8

Given $\triangle ABC$ and a point D such that $A–C–D$, find $m\angle BCD$ if $m\angle A = 103°$ and $m\angle B = 76°$.

Solution

By Theorem 3.12, m∠*BCD* = m∠*A* + m∠*B* = 179°.

∎

EXERCISE SET 3.3

Figure for Exercises 1–5

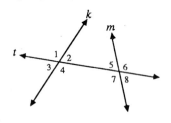

1. List all the interior angles.

2. List all the exterior angles.

3. List all pairs of alternate interior angles.

4. List all pairs of alternate exterior angles.

5. List all pairs of corresponding angles.

6. List all pairs of alternate interior angles.

7. List all pairs of alternate exterior angles.

8. List all pairs of corresponding angles.

Figure for Exercises 6–8

In Exercises 9–10, *u* ‖ *v*.

Figure for Exercises 9–10

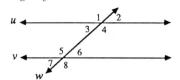

9. If m∠1 = 137°, find the measures of the other angles.

10. If m∠6 = 89°, find the measures of the other angles.

In Exercises 11–14, *a* ‖ *b*.

Figure for Exercises 11–14

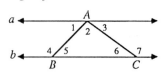

11. *Prove:* m∠1 + m∠3 = m∠5 + m∠6

12. *Prove:* m∠1 + m∠4 = 180°

13. *Prove:* m∠7 > m∠1

14. *Prove:* m∠2 + m∠5 + m∠6 = 180°

In Exercises 15–18, *g* ‖ *h*, *h* ‖ *i*.

Figure for Exercises 15–18

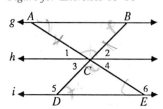

15. *Prove:* m∠5 = m∠1 + m∠*ACB*

16. *Given:* *AC* = *BC*
    Prove: m∠*CDE* = m∠*BAC*

17. *Given:* $AC = BC$
Prove: $m\angle 5 = m\angle 6$

18. *Given:* $AC \neq BC$
Prove: $m\angle 3 \neq m\angle 4$

Figure for Exercises 19–23

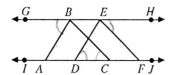

In Exercises 19–23, $\overleftrightarrow{GH} \parallel \overleftrightarrow{IJ}$.

19. *Given:* $m\angle GBA = m\angle EDC$
Prove: $\overleftrightarrow{AB} \parallel \overleftrightarrow{DE}$

20. *Given:* $m\angle HEF = m\angle ACB$
Prove: $\overline{BC} \parallel \overline{EF}$

21. *Given:* $m\angle IAB + m\angle BED = 180°$
Prove: $\overleftrightarrow{AB} \parallel \overleftrightarrow{DE}$

22. *Given:* $\triangle ABC \simeq \triangle DEF$
Prove: $\overline{AB} \parallel \overline{DE}$ and $\overline{BC} \parallel \overline{EF}$

23. *Given:* $\overleftrightarrow{AB} \cap \overleftrightarrow{DE} \neq \emptyset$
Prove: $\triangle ABC \not\simeq \triangle DEF$

Figure for Exercise 24

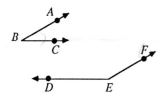

24. *Prove:* If $\overrightarrow{BA} \parallel \overrightarrow{ED}$ and $\overrightarrow{BC} \parallel \overrightarrow{EF}$, then $m\angle B = m\angle E$.
(Hint: Extend $\overrightarrow{BC}$ and $\overrightarrow{ED}$ so as to intersect at P.)

25. *Prove:* If $\overrightarrow{BA} \parallel \overrightarrow{EF}$ and $\overrightarrow{BC} \parallel \overrightarrow{ED}$, then $m\angle B + m\angle E = 180°$.

Figure for Exercise 25

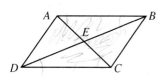

In Exercises 26–27, $A - E - C$ and $B - E - D$.

26. *Prove:* $\triangle ABE \simeq \triangle CDE$ if and only if $\triangle AED \simeq \triangle CEB$

27. *Prove:* $\triangle ABD \simeq \triangle CDB$ if and only if $\triangle ABC \simeq \triangle CDA$

28. Prove Theorem 3.5.

29. Finish the proof of Theorem 3.7.

30. Prove Theorem 3.8, Part (b).

31. Prove Theorem 3.9.

32. Prove Theorem 3.10.

33. If in $\triangle ABC$, $m\angle A = 75°$ and $m\angle B = 2(m\angle C)$, find $m\angle B$ and $m\angle C$.

Figure for Exercises 26–27

Figure for Exercise 34

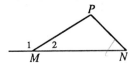

34. *Given:* △*MNP* with exterior ∠1,
     m∠1 = 160°, m∠*N* = 85°
   Find: m∠*P* and m∠2. Give reasons.

In Exercises 35–38, given: m∠A = 75°, m ∠B = 60°, and m∠D = 35°.

35. Find m∠*ACB* and explain.

Figure for Exercises 35–38

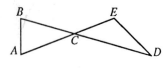

36. Find m∠*E* and explain.

37. Find m∠*BCE* and explain.

38. Find the relation between m∠*A*, m∠*B*, and m∠*BCE*.

39. *Given:* m∠*ABC* = 90°,
      m∠*CBD* = 2(m∠*C*),
      m∠*BFA* = 120°,
      m∠*DBE* = m∠*A*,
      AF = *BF*.

Figure for Exercise 39

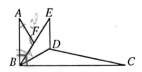

   Find: m∠*A*, m∠*ABF*,
      m∠*DBC*, and m∠*C*.

40. Draw a line *k* and a point *P* not on *k*. Construct the line through *P* parallel to *k* using the method shown in this section.

41. Do Construction 3.1 using corresponding angles.

42. Do Construction 3.1 using alternate exterior angles.

In Exercises 43–44, given: $\overline{AE} \parallel \overline{BD}$, m∠C = 30°, m∠CBD = 40°.

Figure for Exercises 43–44

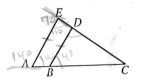

43. *Prove:* m∠*E* = 110°

44. *Prove:* *AC* > *CE*   SIDES oPPi by R Lye,

45. *Given:* *C* is the midpoint
      of $\overline{BD}$, m∠*A* = m∠*E*

Figure for Exercise 45

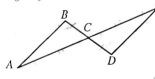

   Prove: △*ABC* ≃ △*EDC*

46. *Given:* *k*∥*m*.

Figure for Exercise 46

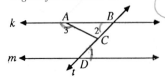

   Prove: m∠3 = m∠1+ m∠2

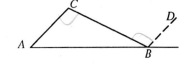

47. Prove Theorem 3.12. (Hint: Construct $\overline{BD} \parallel \overline{AC}$.)

48. Prove Congruence Theorem SAA = SAA.

3.4 POLYGONS

We now generalize the concept of a triangle by considering geometric objects that may have more than three sides. See Figure 3.7.

Figure 3.7

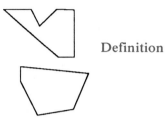

Definition | A **polygon** is the union of n consecutive segments in a plane, intersecting at and only at their endpoints, such that exactly two segments contain each endpoint and no two consecutive segments are on the same line. Each segment is a **side** and each endpoint is a **vertex** of the polygon.

Although there are many types of polygons, in this text we are mainly interested in *convex* polygons. *Rules don't work on concave polygons*

Definition | A polygon is **convex** if for all pairs of points A and B on distinct sides of the polygon, all points between A and B are contained in the interior of the polygon.

Figure 3.8

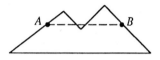

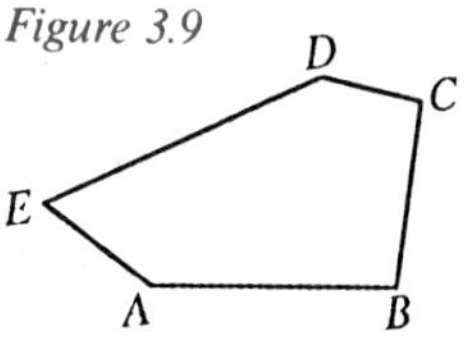

The polygon in Figure 3.8 is not convex, since the points between A and B are not all in the interior of the polygon. The polygon in Figure 3.9 is convex. In the remainder of this text, when we say *polygon* we mean *convex polygon*.

A polygon is **regular** if all its angles and all its sides are congruent. A **diagonal** of a polygon is a line segment joining two nonconsecutive vertices. In Figure 3.9, one of the diagonals is the segment joining the points A and D, namely $\overline{AD}$. The **perimeter** of a polygon is the sum of the lengths of its sides.

Figure 3.9

Example 1

Name all the diagonals of the polygon in Figure 3.9.

Solution

The diagonals are: $\overline{AD}, \overline{AC}, \overline{BD}, \overline{BE}, \overline{CE}.$

Example 2

Find the perimeter of the polygon in Figure 3.9, given that $AB = BC = 7$, $CD = 3$, $DE = 10$, and $AE = 5$.

Solution

The perimeter $= 7 + 7 + 3 + 10 + 5 = 32.$

Many polygons have special names, depending on the number of their sides. See the following table.

Polygon	Number of sides
triangle	3
quadrilateral	4
pentagon	5
hexagon	6
heptagon	7
octagon	8
nonagon	9
decagon	10
n-gon	n

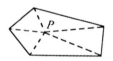

Theorem 3.13   The sum of the measures of the interior angles of an n-gon is $(n-2)180°$.

Proof suggestion: Take a point P in the interior of the n-gon and draw line segments from P to each of the vertices of the n-gon. This forms n triangles. By Theorem 3.11, the sum of the measures of the angles of n triangles must be $n(180)°$. However, the angles that have a vertex P are not interior angles of the n-gon. The sum of the measures of these angles is $360°$. Note that $180n - 360 = (n-2)180$.

◼

Example 3

Find the sum of the measures of the interior angles of a pentagon.

Solution

The sum $= (5-2)180° = 3(180)° = 540°$.

◼

◼

Example 4

If the sum of the measures of the interior angles of a polygon is $900°$, how many sides does the polygon have?

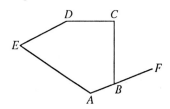

Diagonals $\left(\dfrac{N(N-3)}{2}\right)$

Solution

The sum $= 900° = (n - 2)180°$. Dividing both sides of the equation by $180°$ yields $n - 2 = 5$, so that $n = 7$.

∎

Figure 3.10

If A, B, and C are three consecutive vertices of a polygon and F is a point such that $A-B-F$, then $\angle CBF$ is an **exterior angle** of the polygon. See Figure 3.10.

The proof of the next theorem is left as an exercise.

Theorem 3.14	The sum of the measures of the exterior angles of a polygon, one at each vertex, is $360°$.

∎

Example 5

Find the sum of the measures of the exterior angles of a 132-gon.

Solution

The sum $= 360°$.

∎

EXERCISE SET 3.4

In Exercises 1–4, how many distinct diagonals does each of the polygons have?

1. A triangle

2. A quadrilateral

3. A pentagon

4. An n-gon

In Exercises 5–8, find the sum of the measures of the interior angles of each of the polygons.

5. A triangle

6. A quadrilateral

7. A hexagon

8. An octagon

In Exercises 9–12, find the number of sides of the polygon for which the sum of the measures of the interior angles are given as shown.

  9. 1260°   **11.** 2880°

10. 1440°   **12.** 3780°

In Exercises 13–14, show that it is impossible to have a polygon for which the sum of the measures of the interior angles are given as shown.

13. 1720°   **14.** 2060°

15. Form a table of regular polygons for $n = 3, 4, 5, \ldots, 10$, and for $n = k$. Use the following column headings: n, sum of measures of int. $\angle$s, measure of each int. $\angle$, measure of each ext. $\angle$, and sum of measures of ext. $\angle$s.

Figure for Exercise 16

16. Prove Theorem 3.14. (First prove the theorem for a pentagon, then prove the general case.)

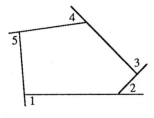

17. A shoe rack in the shape of a quadrilateral is unstable. Explain why a metal brace along a diagonal of the quadrilateral will make the shoe rack stable. Why will a piece of rope in the same position not make the shoe rack stable?

18. A bicycle rack in the shape of a pentagon is unstable. Explain why a metal brace along one diagonal of the pentagon will not make the bicycle rack stable.

3.5   PARALLELOGRAMS

We now discuss a specific type of quadrilateral.

Definition	A **parallelogram** ($\square$) is a quadrilateral in which both pairs of opposite sides are parallel.

Theorem 3.15	A diagonal divides a parallelogram into two congruent triangles.

Given: $\overline{AD} \parallel \overline{BC}$, $\overline{AB} \parallel \overline{CD}$

Prove: $\triangle ACD \simeq \triangle CAB$

Proof:

Statements	Reasons
1. $\overline{AD} \parallel \overline{BC}$	1. Given
2. $m\angle 2 = m\angle 3$	2. If 2 ‖ lines are cut by a trans., the alt. int. $\angle$s are $\simeq$
3. $\overline{AB} \parallel \overline{CD}$	3. Given
4. $m\angle 1 = m\angle 4$	4. Same as 2
5. $AC = AC$	5. Reflexivity
6. $\triangle ACD \simeq \triangle CAB$	6. ASA = ASA

Corollary

> The opposite sides and opposite angles of a parallelogram are congruent.

Given: $\overline{AD} \parallel \overline{BC}$, $\overline{AB} \parallel \overline{CD}$

Prove: $\angle DAB \simeq \angle BCD$, $\angle ABC \simeq \angle CDA$, $\overline{AD} \simeq \overline{BC}$, $\overline{AB} \simeq \overline{CD}$

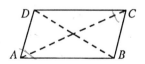

Proof:

Statements	Reasons
1. Construct $\overline{AC}$ and $\overline{BD}$	1. Two pts. determine a line
2. $\overline{AD} \parallel \overline{BC}$, $\overline{AB} \parallel \overline{CD}$	2. Given
3. $\triangle ACD \simeq \triangle CAB$, $\triangle ABD \simeq \triangle CDB$	3. A diag. divides a ▱ into 2 $\simeq \triangle$s
4. $\angle DAB \simeq \angle BCD$, $\angle ABC \simeq \angle CDA$, $\overline{AD} \simeq \overline{BC}$, $\overline{AB} \simeq \overline{CD}$	4. Corr. parts of $\simeq \triangle$s are $\simeq$

The proofs of the following theorems are left as exercises.

Theorem 3.16

> If the opposite angles or the opposite sides of a quadrilateral are congruent, the quadrilateral is a parallelogram.

| Theorem 3.17 | Two parallel lines are everywhere equidistant. |

| Theorem 3.18 | The diagonals of a quadrilateral bisect each other if and only if the quadrilateral is a parallelogram. |

| Theorem 3.19 | Two consecutive angles of a parallelogram are supplementary. |

The converse of Theorem 3.19 is false. (Why?)

A **rhombus** is a parallelogram with a pair of congruent adjacent sides. A **rectangle** (□) is a parallelogram with a right angle. A **square** is a rhombus that is a rectangle.

Example 1

If m∠A = 30° in ▱ABCD, find m∠B, m∠C, and m∠D.

Solution

m∠C = 30° since opposite angles of a parallelogram are congruent.
m∠B = m∠D = 150° since two consecutive angles of a parallelogram are supplementary.

■

Example 2

Can a rectangle be a square? Why?

Solution

Yes. A rectangle is a square if the rectangle has a pair of congruent adjacent sides.

■

Example 3

In quadrilateral $ABCD$, if $m\angle ACD = m\angle CAB$,
(a) Is $\overline{AB} \parallel \overline{CD}$?
(b) Is $\overline{AD} \parallel \overline{BC}$?

Solution

(a) Yes, $\overline{AB} \parallel \overline{CD}$, because if two lines are cut by a transversal so that alternate interior angles are congruent, then the lines are parallel. In this situation, $\angle ACD$ and $\angle CAB$ are alternate interior angles cut by the transversal $\overline{AC}$.

(b) No, $\overline{AD}$ is not necessarily parallel to $\overline{BC}$. There is not enough information given to determine whether they are parallel.

In the study of physics, a quantity that has both magnitude and direction is called a **vector**. Some examples of vectors are velocity, force, and displacement. A problem usually involves more than one vector, and the vectors often must be added to find the **resultant** vector.

A vector is represented in a diagram by a line segment with an arrowhead to show direction. The length of the line segment relative to some number scale represents the magnitude of the vector. The resultant of two vectors is found by the **parallelogram law**. If the directed segments that represent two vectors are drawn from the same initial point, and a parallelogram is constructed with these segments as adjacent sides, then the diagonal drawn from the same initial point represents the resultant of the two given vectors.

Example 4

Figure 3.11

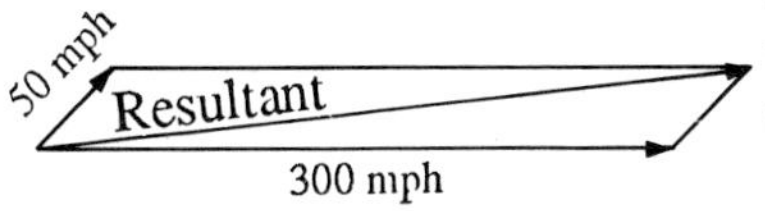

If a plane is flying at a constant velocity of 300 miles per hour (airspeed) due east and a wind of 50 miles per hour is blowing from the southwest, use a diagram to show the vector that represents the resultant velocity, the ground speed, and the direction of flight.

Solution

In Figure 3.11, the resultant velocity is shown by the diagonal of the parallelogram. The length of the diagonal represents ground speed and its direction represents the actual course of the plane.

EXERCISE SET 3.5

Figure for Exercise 2

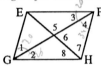

1. If $m\angle A = 50°$ in $\square ABCD$, find $m\angle B$, $m\angle C$, and $m\angle D$. Give reasons.

2. If $m\angle 1 = 30°$, and $m\angle 2 = 20°$, find $m\angle 3$, $m\angle 4$, $m\angle 5$, $m\angle 6$, $m\angle 7$, and $m\angle 8$. Give reasons.

3. Is the set of all squares a subset of the set of all rectangles? Why?

4. Is the set of all squares a subset of the set of all parallelograms? Why?

5. Can a rhombus be a rectangle? Why?

6. If two distinct congruent isosceles triangles have a common base, what special type of parallelogram do they form?

In Exercises 7–9, *KLMN* is a parallelogram. Give a reason for each of the statements.

Figure for Exercises 7–9

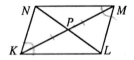

7. $\triangle KLM \cong \triangle MNK$ Def llgram

8. $\overline{KP} \cong \overline{PM}$ Diag of llgram

9. $\angle NKL \cong \angle LMN$ Def llgram

10. If $AB = 2$ inches, $AD = 1.5$ inches, and $m\angle DAB = 55°$, construct $\square ABCD$.

11. Construct $\square MNPQ$ if $MN = 2.5$ inches and $MP = 1$ inch.

12. Construct a rhombus with a given side $\overline{PQ}$.

13. Construct a square with a given side $\overline{MN}$.

14. *Given:* $\angle A$, $AB = 3.5$ inches, $AD = 2$ inches.

    Construct: $\square ABCD$ as follows: Copy $\angle A$ on a working line and copy $\overline{AB}$ and $\overline{AD}$ on the sides of $\angle A$. Then, with B as center, make an arc of radius AD, and with D as center make an arc of radius AB. The intersection of the arcs is the vertex C of the $\square$. Why? Can you think of another method of construction?

In Exercises 15–19, choose a suitable scale and make a construction to illustrate each vector problem.

15. A plane is heading due west at 360 miles per hour and a 45 mile per hour wind is blowing from the southeast. Find the resultant course and ground speed.

16. Two forces of 40 pounds and 60 pounds act upon an object at an angle of 70° to each other. Find the resultant force (the magnitude and the angle between the resultant and the 40 pound force).

17. Two forces of 30 pounds and 40 pounds act upon an object at a 90° angle to each other. Find the magnitude of the resultant. What relationship is there between the magnitudes of the three forces?

18. The two vectors used to find a resultant are *components*. Find the vertical and horizontal components of a 70 pound force if it makes an angle of 30° with the horizontal.

19. Find the resultant of three vectors of magnitude 3, 5, and 6. (Hint: Add any two of the vectors and then add the third vector to the resultant of the first two.)

20. Prove that if two sides of a quadrilateral are congruent and parallel, then the quadrilateral is a parallelogram.

21. Prove Theorem 3.16.

22. Prove Theorem 3.17, first stating it in "if-then" form.

23. Prove Theorem 3.18, first stating it in "if-then" form.

24. Prove Theorem 3.19, first stating it in "if-then" form.

3.6 INTERCEPTS

The proof of the next theorem is left as an exercise.

| Theorem 3.20 | Two distinct lines parallel to a third line are parallel to each other. |

If two lines intersect a third line at distinct points A and B, then the two lines **intercept** $\overline{AB}$.

| Theorem 3.21 | If three or more parallel lines intercept congruent segments on one transversal, they intercept congruent segments on every transversal. |

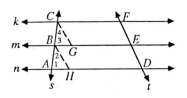

Given: $k \parallel m \parallel n$, AB = BC

Prove: $EF = DE$

Proof:

Statements	Reasons
1. Construct $\overline{CG} \parallel \overline{EF}$, $\overline{BH} \parallel \overline{ED}$	1. Parallel Post.
2. $\overline{CG} \parallel \overline{BH}$	2. Two distinct lines $\parallel$ to a 3rd line are $\parallel$ to each other
3. $\angle 2 \simeq \angle 4$	3. If 2 $\parallel$ lines are cut by a trans., corr. $\angle$s are $\simeq$
4. $m \parallel n$, $k \parallel m$, AB = BC	4. Given
5. $\angle 1 \simeq \angle 3$	5. Same as 3
6. $\triangle ABH \simeq \triangle BCG$	6. ASA = ASA
7. $CG = BH$	7. Corr. parts of $\simeq \triangle$s are $\simeq$
8. quadrilaterals *BEDH* and *CFEG* are $\square$s	8. Def. of $\square$
9. $EF = CG$, $BH = DE$	9. Opp. sides of a $\square$ are $\simeq$
10. $EF = DE$	10. Transitivity (State. 7 and 9)

The proof of the next theorem is left as an exercise.

Theorem 3.22	If a segment joins the midpoints of two sides of a triangle, then it is parallel to the third side and has half the length of the third side.

Example 1

Given $\triangle ABC$ with $AC = 20$ and points D and E the midpoints of segments $\overline{AB}$ and $\overline{BC}$, respectively, find DE.

Solution

By Theorem 3.22, $DE = 10$.

Example 2

Given $\triangle ABC$ with points D and E the midpoints of segments $\overline{AC}$ and $\overline{BC}$, respectively; if $DE = 6$, find AB.

Solution

By Theorem 3.22, $AB = 12$.

Definition | A **trapezoid** is a quadrilateral with exactly one pair of parallel sides. The parallel sides are *bases*. The nonparallel sides are *legs*.

Figure 3.12

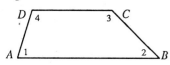

A pair of angles of a trapezoid are called *base angles* if they both contain the same base. The line segment joining the midpoints of the legs is the *median* of the trapezoid. An **isosceles trapezoid** is one in which the nonparallel sides are congruent. In Figure 3.12, let $\overline{AB} \parallel \overline{CD}$. Sides $\overline{AB}$ and $\overline{CD}$ are bases. Angles 3 and 4 are a pair of base angles. Angles 1 and 2 are another pair of base angles.

The proofs of the following two theorems are left as exercises.

Theorem 3.23 | Base angles of an isosceles trapezoid are congruent.

Example 3

Given isosceles trapezoid $EFGH$ with bases $\overline{EF}$ and $\overline{GH}$, find $m\angle E$, $m\angle F$, and $m\angle G$ if $m\angle H = 30°$.

Solution

By Theorem 3.23, $m\angle G = 30°$. By Theorem 3.13, the sum of the measures of the interior angles of a quadrilateral is $360°$. Thus,

$$m\angle E + m\angle F = 360° - 2(30°) = 300°.$$

Therefore, by Theorem 3.23, $m\angle E = m\angle F = 150°$.

Theorem 3.24	The median of a trapezoid is parallel to both bases and its length is equal to half the sum of the lengths of the bases.

■

Example 4

Find the length of the median of the trapezoid shown in Figure 3.12 if $AB = 17$ and $CD = 5$.

Solution

By Theorem 3.24, the median has length 11.

■

Construction 3.2	Divide a line segment into a given number of congruent segments.

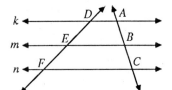

Given: $\overline{AB}$

Construct: C, D, E such that $\overline{AC} \simeq \overline{CD} \simeq \overline{DE} \simeq \overline{EB}$

Step 1. For any point P not on $\overleftrightarrow{AB}$, construct $\overrightarrow{AP}$.

Step 2. Choose any convenient length and construct four congruent segments $\overline{AR}, \overline{RS}, \overline{ST}$, and $\overline{TU}$ on $\overrightarrow{AP}$. Construct $\overline{BU}$.

Step 3. Copy $\angle TUB$ at R, S, and T. Label the intersections of these angles with $\overline{AB}$ by C, D, and E, respectively. Then $\overline{CR} \parallel \overline{DS} \parallel \overline{ET} \parallel \overline{BU}$ by Theorem 3.8 and, hence, $\overline{AC} \simeq \overline{CD} \simeq \overline{DE} \simeq \overline{EB}$ by Theorem 3.21.

EXERCISE SET 3.6

Figure for Exercise 1

1. If $k \parallel m \parallel n$, $AB = BC = 2$, and $DE = 3$, find EF.

2. Find the length of the median of trapezoid $ABCD$ if $\overline{BC}$ and $\overline{DA}$ are its bases, $BC = 5$, and $DA = 8$.

Figure for Exercise 3

3. If *ABCD* is an isosceles trapezoid with bases $\overline{AB}$ and $\overline{CD}$ and m∠*C* = 100°, find m∠*A*, m∠*B*, and m∠*D*. Give reasons.

In Exercises 4–7, let *k* ∥ *m* ∥ *n*. Using the values given on the figure, find the numerical values of each of the following.

4. *BE*

5. *BC*

6. m∠1

7. m∠2

Figure for Exercises 4–7

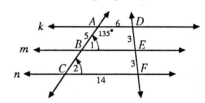

In Exercises 8–10, △*ABC* is given with m∠*C* = 100° and $\overline{AM} \simeq \overline{MC}$. Give a reason for each of the statements.

8. m∠*B* < 90°.

9. If $\overline{CN} \simeq \overline{NB}$, then $\overline{MN}$ ∥ $\overline{AB}$.

10. If m∠*A* = 30°, then m∠*B* = 50°.

11. By construction, divide a given line segment $\overline{AB}$ into three congruent segments.

12. By construction, divide a given line segment $\overline{AB}$ into five congruent segments.

13. Prove that the quadrilateral *MNPQ* formed by joining the consecutive midpoints of any quadrilateral *ABCD* is a ▱. (Hint: Draw $\overline{BD}$ and consider △*ABD* and △*BCD* and Theorem 3.22.)

14. Prove Theorem 3.20.

15. Prove Theorem 3.22. (Hint: Construct $\overline{FG}$ on $\overleftrightarrow{DF}$ such that *FG* = *DF*. Construct $\overline{BG}$. Show that quadrilateral *ABGD* is a parallelogram.)

16. Prove Theorem 3.23. (Hint: Construct $\overline{CP}$ ∥ $\overline{AD}$. Show that quadrilateral *APCD* is a parallelogram.)

17. Prove Theorem 3.24. (Hint: Construct $\overrightarrow{CE}$ and $\overrightarrow{BA}$ intersecting in point *P*. Show that *PE* = *CE* by proving that △*APE* ≃ △*DCE*. Consider △*PCB* and use Theorem 3.22.)

18. Prove that a line bisecting one side of a triangle and parallel to a second side bisects the third side.

Figure for Exercises 8–10

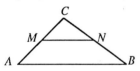

Figure for Exercise 13

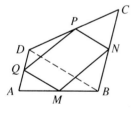

Figure for Exercise 16

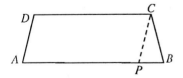

Figure for Exercise 17

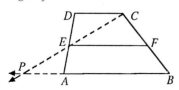

Figure for Exercise 15

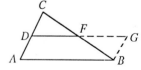

Chapter 3
Key Terms

Chapter 3
Review Exercises

Fill each blank with the word *always, sometimes,* or *never.*

1. The sum of the measures of the angles of a triangle is _____ 180°.

2. The diagonals of a quadrilateral _____ divide it into four congruent triangles.

3. The diagonals of a trapezoid are _____ perpendicular.

4. A trapezoid is _____ a parallelogram.

5. A square is _____ a rhombus.

6. If two lines are cut by a transversal, the alternate interior angles formed are _____ equal.

7. An exterior angle at the base of an isosceles triangle is _____ an obtuse angle.

8. A line that joins the midpoints of two sides of a triangle is _____ parallel to the third side.

9. If the number of sides of a polygon is doubled, the sum of the measures of the exterior angles of this polygon _____ changes.

10. A diagonal of a rhombus is _____ congruent to a side.

True-False: If the statement is true, mark it so. If it is false, replace the underlined word to make a true statement.

11. Two consecutive angles of a parallelogram are <u>complementary.</u>

12. A <u>rhombus</u> is a quadrilateral with two and only two parallel sides.

13. The measure of an exterior angle of a triangle is <u>equal to</u> the sum of the measures of the remote interior angles.

14. If one angle of a triangle is a right angle, the other two angles are <u>supplementary.</u>

15. The opposite sides of any <u>parallelogram</u> are congruent.

16. The diagonals of a parallelogram are <u>perpendicular</u> to each other.

17. The sum of the measures of the interior angles of a <u>hexagon</u> is 1080°.

18. Two <u>opposite</u> angles of a parallelogram are supplementary.

19. The perimeter of the triangle formed by connecting consecutive midpoints of the sides of $\triangle ABC$ is <u>one-half</u> the perimeter of $\triangle ABC$.

20. The diagonals of an isosceles trapezoid are <u>congruent</u>.

Answer the following questions.

21. Two parallel lines are cut by a transversal. Two interior angles on the same side of the transversal are represented by x and $2x$. Find the measure of each angle.

22. Two parallel lines are cut by a transversal. Two exterior angles on the same side of the transversal are represented by y and $2y + 30°$. Find the measure of each angle.

23. The sides of one angle are parallel to the sides of another. If one angle is 50°, what are the possible values of the other angle? Explain.

24. Find the vertex angle of an isosceles triangle if one of the base angles has a measure of 37°.

25. Find each base angle of an isosceles triangle if each base angle is twice the vertex angle.

26. An exterior angle at the base of an isosceles triangle is 155°. What is the measure of the vertex angle?

27. In $\triangle ABC$, $m\angle A = 5x$, $m\angle B = 7x$, and $m\angle C = 36°$. Find the measures of $\angle A$ and $\angle B$.

28. If one angle of a triangle is twice the smallest angle and the third angle is three times the smallest angle, find the measures of each angle.

29. The sum of the interior angles of a polygon is 1800°. How many sides does the polygon have?

30. Find the measure of each exterior angle of a regular octagon.

31. How many sides does a polygon have if the sum of the measures of its exterior angles equals the sum of the measures of its interior angles?

32. What is the sum of the measures of the interior angles of a 61-gon? What is the sum of the measures of the exterior angles?

33. In $\square ABCD$, $m\angle B$ is twice as large as $m\angle A$. Find $m\angle B$.

34. In quadrilateral $ABCD$, $AB = CD$, $BC = AD$, $m\angle DCA = 55°$. Find $m\angle CAB$.

35. If the midpoint of side $\overline{AB}$ of $\triangle ABC$ is D, and a line from D parallel to $\overline{BC}$ cuts $\overline{AC}$ at P, and if $DP = 12$, find BC.

36. If $\overline{AB} \parallel \overline{CD}$, find the measures of all the angles when $m\angle 1 = 64°$ in the figure.

Figure for Exercise 36

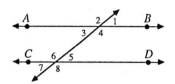

Proofs:

37. Prove the diagonals of a square are perpendicular to each other.

38. Prove the diagonals of a rhombus are perpendicular to each other.

Figure for Exercise 40

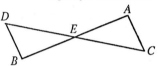

39. In quadrilateral $ABCD$, $AB = BC$ and $AD = CD$. Prove $\overline{BD} \perp \overline{AC}$.

40. In the figure, prove $\overline{CA} \parallel \overline{BD}$ if $\overline{AB}$ and $\overline{CD}$ bisect each other.

41. In the figure, if $EC = BF$, $AB = CD$, $\overline{FB} \perp \overline{AD}$, and $\overline{EC} \perp \overline{AD}$, prove $\overline{EA} \parallel \overline{DF}$.

Figure for Exercise 41

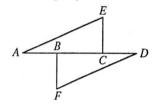

42. In the figure, prove $\overline{AB} \parallel \overline{CD}$ if $m\angle 1 = 147°$ and $m\angle 2 = 33°$.

43. Prove that if one angle of a triangle has a measure equal to the sum of the measures of the other two, then the triangle is a right triangle.

Figure for Exercise 42

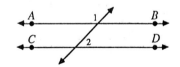

44. Prove that in a parallelogram, perpendiculars to a diagonal from two opposite vertices have the same length.

45. *Given:* $AP = BP$,
$\overline{AD} \perp \overline{CP}$,
$\overline{BE} \perp \overline{CE}$
Prove: $AEBD$ is a parallelogram.

Figure for Exercise 45

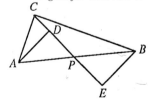

46. Prove that a parallelogram is a rhombus if a diagonal bisects a pair of opposite angles.

47. Prove that the segments joining the midpoints of the sides of a triangle divide the triangle into four congruent triangles.

48. *Given:* $AC = BC$,
$\overline{MP} \perp \overline{AC}$,
$\overline{PN} \perp \overline{BC}$
Prove: $m\angle 1 = m\angle 2$.

Figure for Exercise 48

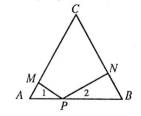

Constructions:

49. Construct a rhombus that has two given line segments as its diagonals. (Hint: The diagonals of a rhombus are perpendicular bisectors of each other.)

50. Construct an equilateral triangle whose perimeter is a given line segment.

In Exercises 51–52, choose a suitable scale and make a construction to illustrate each vector problem.

51. Two forces of 55 pounds and 70 pounds act upon an object at an angle of 35° to each other. Find the resultant force.

52. Two forces of 22 pounds and 40 pounds act upon an object at an angle of 65° to each other. Find the resultant force.

Similar and Regular Polygons

Courtesy of Historical Pictures Service, Chicago

Parthenon

Historical Note

As early as Greek times, it was thought that the most artistic dimensions for a rectangle were those involved in the *golden section* or the *divine proportion*. The length L and width W of such a rectangle satisfy the proportion

$$\frac{L + W}{L} = \frac{L}{W}.$$

It can be shown that if a line segment has length $C = L + W$, then $W = C(3 - \sqrt{5})/2$ and $L = C(\sqrt{5} - 1)/2$. The ratio of length to width is approximately eight to five. From early times, the golden section was used extensively in art and architecture, including the measurements of the Parthenon on the Acropolis of Athens.

...scussed the concept of congruence of triangles, in ...tially have the same shape and size. In this chap- ...ncept of similar polygons, which have the same ...sarily the same size.

...ession of the form a/b, $b \neq 0$, where a and b are real ...io is sometimes written as $a{:}b$ and is read "a is to b."

Example 1

Form the ratio of the two numbers 6 and 8.

Solution

The ratio is 6/8, which reduces to 3/4.

Example 2

Form the ratio of 3 gallons and 9 quarts.

Solution

First change the quantities to common units:

$$3 \text{ gallons} = 12 \text{ quarts}$$

Thus, the ratio is (12 quarts)/(9 quarts) = 4/3.

A **proportion** is a statement that two ratios are equal. Thus, $a/b = c/d$, and $4/6 = 2/3$ are proportions. The first proportion can be written as $a{:}b = c{:}d$ and is read "a is to b as c is to d." In general, we say that a sequence of real numbers a, b, c **is proportional to** the sequence of real numbers r, s, t, if $a/r = b/s = c/t$. This last expression is a **continued proportion** and may contain any number of ratios. Each ratio of a continued proportion is equal to each of the other ratios.

Example 3

Is 1, 2, 3, 4 proportional to 2, 4, 6, 8? Why?

Solution

Yes, since 1/2 = 2/4 = 3/6 = 4/8.

In the proportion $a/b = c/d$, a and d are the **extremes**, b and c are the **means**, and d is the **fourth proportional** to a, b, c. By symmetry, we must also have $c/d = a/b$. However, notice that this changes the means, the extremes, and the fourth proportional.

Example 4

In the proportion $10/17 = 20/34$, name the means, the extremes, and the fourth proportional.

Solution

The means are 17 and 20.
The extremes are 10 and 34.
34 is the fourth proportional to 10, 17, 20.

In the proportion $a/b = b/c$, b is the **geometric mean (mean proportional)** between a and c.

Example 5

Find the mean proportional between 4 and 25.

Solution

We need b such that $4/b = b/25$. This means $b^2 = 100$. Thus, $b = 10$ or -10. (However, in geometry, we are concerned with only the positive value.)

Example 6

Find the fourth proportional to 3, 9, 9.

Solution

We need c such that $3/9 = 9/c$. This means $3c = 81$. Thus, $c = 27$.

Figure 4.1

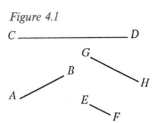

Segments whose lengths are proportional are **proportional segments**. Thus, in Figure 4.1, if $AB = 3$, $CD = 7$, $EF = 3/2$, $GH = 7/2$, then the segments $\overline{AB}$ and $\overline{CD}$ are proportional to the segments $\overline{EF}$ and $\overline{GH}$, since $AB/CD = EF/GH$.

Example 7 is an algebraic manipulation of proportions and a proof of the statement "In any proportion, the product of the means is equal to the product of the extremes."

Example 7

Prove: If $\dfrac{a}{b} = \dfrac{c}{d}$, then $ad = bc$.

Solution

Given $\dfrac{a}{b} = \dfrac{c}{d}$, using the multiplication axiom of equality, multiply both sides by bd to get the result.

EXERCISE SET 4.1

The statements in Exercises 1–6 can be treated as theorems. State each of the theorems in words, and prove them using algebraic axioms.

1. If $\dfrac{a}{b} = \dfrac{c}{d}$, then $\dfrac{a}{c} = \dfrac{b}{d}$.

2. If $\dfrac{a}{b} = \dfrac{c}{d}$, then $\dfrac{b}{a} = \dfrac{d}{c}$.

3. If $\dfrac{a}{b} = \dfrac{b}{c}$, then $b = \pm\sqrt{ac}$.

4. If $\dfrac{a}{b} = \dfrac{c}{d}$, then $\dfrac{a+b}{b} = \dfrac{c+d}{d}$.

5. If $\dfrac{a}{b} = \dfrac{c}{d}$, then $\dfrac{a-b}{b} = \dfrac{c-d}{d}$.

6. If $\dfrac{a}{b} = \dfrac{c}{d}$, then $\dfrac{a}{b} = \dfrac{c}{d} = \dfrac{a+c}{b+d}$.

Express each of Exercises 7–14 as a fraction and simplify.

7. The ratio of 10 to 15   **9.** 72:24

8. The ratio of 16 to 20   **10.** 3/4:9/16

11. The ratio of 15° to 90°

12. The ratio of 12 meters to 15 meters

13. The ratio of 27 inches to 3 feet

14. The ratio of 150′ to 5°

In Exercises 15–20, solve for x in each proportion.

15. $20/65 = x/13$ **18.** $4:(x+1) = 3:x$

16. $x/3 = 4/5$ **19.** $2/3:x = 9:4$

17. $(x-2):8 = 5:2$ **20.** $3x:5x = 6:10$

In Exercises 21–24, find the fourth proportional to:

21. 2, 3, 4   **23.** 5, 8, 4

22. 4, 8, 12   **24.** 1, 100, 10

In Exercises 25–28, find the mean proportional between:

25. 3 and 27   **27.** 1/4 and 9

26. 4 and 16   **28.** 4 and 3

In Exercises 29–32, write each as a continued proportion.

29. 3, 4, 5 is proportional to 9, 12, 15.

30. 1, 3, 4, 7 is proportional to 5, 15, 20, 35.

31. 4, 8, 12, 16 is proportional to 6, 12, 18, 24.

32. 1/2, 1/3, 1/4, 1/6 is proportional to 6, 4, 3, 2.

33. Suppose a soup recipe that serves 110 people requires the following ingredients:

> 50 quarts of water
> 1 cup of salt
> 5 pounds of carrots
> 12 chickens
> 3 pounds of celery

What quantity of each item in this recipe is needed to serve 8 people?

34. Suppose a soup recipe that serves 50 people requires the following ingredients:

> 7 gallons of water
> 1/2 cup of pepper
> 3 pounds of lentils
> 1 turkey
> 3 pounds of onions
> 6 ounces of hot sauce

What quantity of each item in this recipe is needed to serve 300 people?

35. The earth has a diameter of approximately 8,000 miles and is approximately 240,000 miles from the moon, which has a diameter of approximately 2200 miles. A scale drawing of the earth and moon is made, in which the moon is represented by a circle with a diameter of 1 cm. What is the diameter of the earth, and how far is the earth from the moon on this scale drawing?

36. Using the data in Exercise 35, a scale drawing of the earth and moon is made, in which the earth is represented by a circle with a diameter of 1 cm. What is the diameter of the moon, and how far is the earth from the moon on this scale drawing?

4.2 SIMILARITY

Similarity is a concept used to compare the angles and sides of different polygons.

Definition	Two polygons are **similar** if corresponding angles are congruent and the corresponding sides are proportional segments.

Figure 4.2

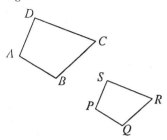

In Figure 4.2, $\angle A \simeq \angle P$, $\angle B \simeq \angle Q$, $\angle C \simeq \angle R$, $\angle D \simeq \angle S$, and $AB/PQ = BC/QR = CD/RS = DA/SP$. Thus, the two quadrilaterals are similar. We write $ABCD \sim PQRS$, using the ordering of letters to indicate the correspondence as we did for congruent triangles.

Similarity is used in applications such as surveying, blueprint making, and the enlargement of photographs. The following example shows how the height of a tree or building may be calculated by using similar triangles.

Example 1

Figure 4.3

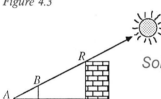

In Figure 4.3, a yardstick $\overline{BC}$ casts a 4-foot shadow $\overline{AC}$ and the shadow $\overline{AT}$ of the building is 50 feet long. Find the height RT of the building.

Solution

$\triangle ABC \sim \triangle ART$. Find RT by solving the proportion $RT/BC = AT/AC$ for RT. We get $RT/3 = 50/4$, so the solution is $RT = 37.5$ feet.

In developing theorems of similarity for triangles, we need the following postulate. (In some logical developments of geometry, this postulate is stated as a theorem and proved, but we will simply assume its validity.)

Postulate
4.1

> **(AAA ~ AAA)** If the three angles of one triangle are congruent to the three angles of another triangle, then the triangles are similar.

Thus, to prove two triangles similar, it is unnecessary to check for proportionality of the corresponding sides, as long as all the angles of one are known to be congruent to the corresponding angles of the other. Theorem 4.1 states that information about only two pairs of angles is sufficient for checking similarity. The proof is left as an exercise.

Theorem
4.1

> **(AA ~ AA)** If two angles of one triangle are congruent to two angles of another triangle, then the triangles are similar.

Example 2

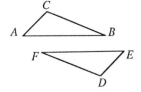

Figure 4.4

Discuss the similarity of the triangles in Figure 4.4, if $\angle A \simeq \angle E$ and $\angle C \simeq \angle D$.

Solution

By Theorem 4.1, $\triangle ABC \sim \triangle EFD$. (The order of the letters in the names for the triangles is important.)

■

The theorem concerning similar triangles can often be used to prove properties of a single triangle. For example, the following theorem is a consequence of Theorem 4.1.

Theorem 4.2	A line parallel to one side of a triangle that intersects the other two sides in distinct points divides these two sides into proportional segments.

Given: $\overline{DE} \parallel \overline{AB}$

Prove: $\dfrac{AD}{DC} = \dfrac{BE}{EC}$

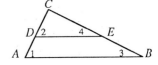

Proof:

Statements	Reasons
1. $\overline{DE} \parallel \overline{AB}$	1. Given
2. $\angle 1 \simeq \angle 2, \angle 3 \simeq \angle 4$	2. Corr. $\angle$s formed by 2 $\parallel$ lines and a trans. are $\simeq$
3. $\triangle CDE \sim \triangle CAB$	3. AA $\sim$ AA
4. $\dfrac{AC}{DC} = \dfrac{BC}{EC}$	4. Def. of $\sim$
5. $AC = AD + DC$ $BC = BE + EC$	5. Def. of between
6. $\dfrac{AD + DC}{DC} = \dfrac{BE + EC}{EC}$	6. Substitution
7. $\dfrac{AD}{DC} + 1 = \dfrac{BE}{EC} + 1$	7. Substitution
8. $\dfrac{AD}{DC} = \dfrac{BE}{EC}$	8. Add. Axiom of $=$

Constructions 4.1 and 4.2 demonstrate a use for Theorems 4.1 and 4.2.

Construction 4.1	Construct the fourth proportional to three given line segments.

Given: $\overline{AB}$, $\overline{CD}$, $\overline{EF}$

Construct: $\overline{UV}$ so that
$$\frac{AB}{CD} = \frac{EF}{UV}$$

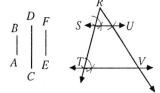

Step 1. Construct an arbitrary angle R. On one ray of $\angle R$, construct $\overline{RS}$ and $\overline{ST}$ so that $RS = AB$, $ST = CD$, and $R-S-T$.

Step 2. On the other ray, construct $\overrightarrow{RU}$ so that $RU = EF$. Draw $\overleftrightarrow{SU}$.

Step 3. Using Construction 3.1, construct a parallel to $\overleftrightarrow{SU}$ through T. This line intersects $\overrightarrow{RU}$ in a point V. The required line segment is $\overline{UV}$, by Theorem 4.2.

Construction 4.2	Construct a polygon similar to a given polygon.

Given: Polygon $ABCDE$, $\overline{GH}$

Construct: Polygon $GHIJK$ so that $GHIJK \sim ABCDE$

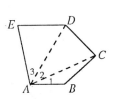 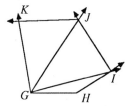

Step 1. Draw diagonals $\overline{AC}$ and $\overline{AD}$ forming angles 1, 2, and 3 at A.

Step 2. Now construct $\triangle GHI \sim \triangle ABC$. Copy $\angle 1$ at G and $\angle B$ at H, labeling their point of intersection I. The triangle formed is similar to $\triangle ABC$ by AA $\sim$ AA.

Step 3. Repeat the construction in Step 2, first constructing $\triangle GIJ$ similar to $\triangle ACD$ and then constructing $\triangle GJK$ similar to $\triangle ADE$. Then, $GHIJK \sim ABCDE$. One may continue this process around a polygon of any number of sides.

EXERCISE SET 4.2

In Exercises 1–3, write the congruences between the corresponding angles and the continued proportion for the corresponding sides.

1. $\triangle ABC \sim \triangle RSQ$

2. $HURT \sim YELP$

3. $KLMNPQ \sim KABCDQ$

4. Find the height of a tree that casts a 90-foot shadow, if an adjacent fence post 2.5 feet high casts a 2-foot shadow.

5. We wish to find the distance AB across the river shown in the diagram. Suppose that $\triangle ABD \sim \triangle ACE$. We find, by direct measurement, that $BD = 60$ feet, $EC = 75$ feet, and $BC = 40$ feet. Find AB.

6. A navigator on a ship wishes to compute the distance from the ship to a buoy. He uses the diagram shown on the left, where point B represents the buoy and points P, Q, S, T are on the ship. Find BQ.

7. Find the height of a building that has a 100-foot shadow, if a man 5 feet tall casts a 7-foot shadow.

8. Find the height of a pole that has a 10-foot shadow, if a woman 6 feet tall casts a 2-foot shadow.

Use the following information for Exercises 9–10: In the study of optics, the focal length c of a lens is given by the equation $1/a + 1/b = 1/c$, where a is the image distance from the lens and b is the object distance from the lens.

Figure for Exercise 5

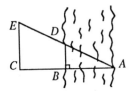

Figure for Exercise 6

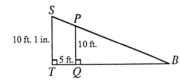

Figure for Exercises 9–10

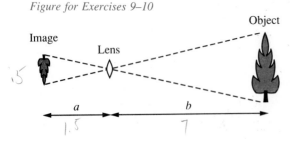

9. If $a = 1.5$ inches, $b = 7$ inches, and the height of the image is 0.5 inches, find the height of the object and the focal length of the lens.

10. If $a = 0.5$ centimeters, $b = 20$ centimeters, and the height of the object is 70 centimeters, find the height of the image and the focal length of the lens.

11. On the blueprint for a building, the scale is 1/16 inch to 1 foot. What are the dimensions of a room that has blueprint measurements of 1.5 inches by 1.25 inches?

12. On the blueprint for a building, the scale is 1/16 inch to 1 foot. If a room has dimensions of 10 feet by 14 feet, what are the corresponding measurements on the blueprint?

In Exercises 13–16, determine whether the two triangles are similar. If so, write the similarity.

13.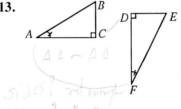

14.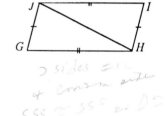

15.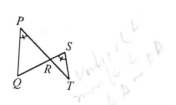

16.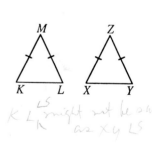

Figure for Exercises 17–18

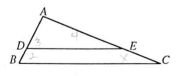

In Exercises 17–18, given: $\triangle ABC$, $\overline{DE} \parallel \overline{BC}$.

17. Find EC if $AD = 3$, $BD = 2$, and $AE = 4$.

18. Find AE if $AE = BD$, $AD = 16$, and $EC = 4$.

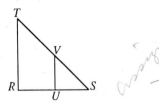

In Exercises 19–21, given $\triangle RST, \overline{UV} \parallel \overline{RT}$

19. Find VT/SV if $SU = 7$ and $UR = 2$.

20. Find UR if $SU = 3, SV = 5$, and $VT = 2$.

21. Find VT if $ST = 12, UR = 8$, and $SU = 10$.

22. Draw line segments so that $AB = 1.5$ inches, $CD = 2$ inches, and $EF = 3$ inches. Construct the fourth proportional to this set of line segments.

23. Draw line segments, $\overline{GH}, \overline{KL}$, and $\overline{MN}$ of various lengths. Construct $\overline{PQ}$ so that $GH{:}KL = PQ{:}MN$.

24. Construct a line segment of length ab, given line segments of length a, b, and $PQ = 1$ unit. (Hint: Since $1/a = b/x$ implies $x = ab$, construct the fourth proportional to 1, a, and b.)

25. By construction, divide $\overline{AB}$ into two segments whose lengths have a ratio of 2 to 3.

26. By construction, divide $\overline{ST}$ into segments proportional to 1 inch, 2 inches, and 1.25 inches, if $ST = 5.5$ inches.

27. Construct a quadrilateral similar to a given quadrilateral.

28. Construct a heptagon similar to a given heptagon.

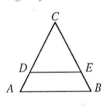

29. *Given:* $A - D - C, B - E - C,$
    $\quad\quad\quad CD = CE, AD = BE$
    Prove: $\triangle CDE \sim \triangle CAB$ and $CD/CA = CE/CB$

30. *Given:* $A - D - C, B - E - C,$
    $\quad\quad\quad \overline{DE} \parallel \overline{AB}$
    Prove: $AB \cdot CD = AC \cdot DE$

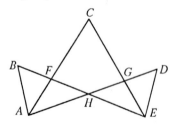

31. *Given:* $\angle CEB \cong \angle ABE,$
    $\quad\quad\quad A - H - G$
    Prove: $\triangle HEG \cong \triangle HBA$

32. *Given:* $A - F - C, E - G - C,$
    $\quad\quad\quad CF = CG, FH = GH,$
    $\quad\quad\quad \angle ABE \cong \angle EDA$
    Prove: $\triangle ABF \sim \triangle EDG$

33. Prove that $\triangle ABC \sim \triangle ABC$. (Reflexivity).

34. Prove that if $\triangle ABC \sim \triangle DEF$, then $\triangle DEF \sim \triangle ABC$. (Symmetry)

35. Prove that if $\triangle ABC \sim \triangle DEF$, and $\triangle DEF \sim \triangle GHI$, then $\triangle ABC \sim \triangle GHI$. (Transitivity)

36. If two polygons are similar, prove that the ratio of their perimeters equals the ratio of the lengths of any two corresponding sides.

37. Prove Theorem 4.1.

4.3 MORE THEOREMS ON SIMILARITY

Theorem 4.3 is a converse of Theorem 4.2. Its proof again illustrates the addition of an auxiliary line to a diagram.

Theorem 4.3

| A line that intersects two sides of a triangle in distinct points and divides these two sides into proportional segments is parallel to the third side. |

Given: $\dfrac{BE}{EC} = \dfrac{AD}{DC}$

Prove: $\overline{AB} \parallel \overline{DE}$

Proof:

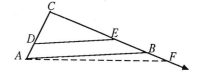

Statements	Reasons
1. Let F be a point on $\overrightarrow{CB}$ such that $\overline{AF} \parallel \overline{DE}$. (Notice that C–E–F)	**1.** Parallel Post.
2. $\dfrac{AD}{DC} = \dfrac{FE}{EC}$	**2.** A line $\parallel$ to one side of a $\triangle$ that intersects the other 2 sides in distinct pts. divides these 2 sides into proportional segments
3. $\dfrac{BE}{EC} = \dfrac{AD}{DC}$	**3.** Given
4. $\dfrac{BE}{EC} = \dfrac{FE}{EC}$	**4.** Transitivity
5. $BE = FE$	**5.** Mult. Axiom of $=$
6. $B = F$	**6.** Def. of length and Ruler Post.
7. $\overline{AB} \parallel \overline{DE}$	**7.** Substitution

Theorem 4.3 is used to prove the following similarity theorem.

<table>
<tr><td>Theorem
4.4</td><td>**(SAS ~ SAS)** If two pairs of corresponding sides of two triangles are proportional and the included angles are congruent, then the two triangles are similar.</td></tr>
</table>

Given: $\dfrac{AC}{DF} = \dfrac{AB}{DE}$, $\angle A \simeq \angle D$

Prove: $\triangle ABC \sim \triangle DEF$

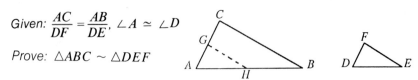

Note: In this proof we assume, without loss of generality, that $DF < AC$, since the proof is similar to the one below, if $DF > AC$, and the proof is trivial, if $DF = AC$ (for then $AB = DE$ and $\triangle ABC \simeq \triangle DEF$).

Proof:

Statements	Reasons
1. Let G be the point of $\overline{AC}$ such that $AG = DF$. Let H be the point of $\overline{AB}$ such that $AH = DE$.	1. Ruler Post.
2. $\angle A \simeq \angle D$	2. Given
3. $\triangle AHG \simeq \triangle DEF$	3. SAS = SAS
4. $\angle F \simeq \angle AGH$, $\angle E \simeq \angle AHG$	4. Corr. parts of $\simeq \triangle$s are $\simeq$
5. $\dfrac{AC}{DF} = \dfrac{AB}{DE}$	5. Given
6. $\dfrac{AC}{AG} = \dfrac{AB}{AH}$	6. Substitution
7. $\overline{GH} \parallel \overline{BC}$	7. A line that intersects 2 sides of a $\triangle$ in distinct points and divides these 2 sides into proportional segments is $\parallel$ to the third side
8. $\angle AGH \simeq \angle C$, $\angle AHG \simeq \angle B$	8. Corr. $\angle$s formed by 2 $\parallel$ lines and a trans. are $\simeq$
9. $\angle F \simeq \angle C$, $\angle E \simeq \angle B$	9. Transitivity (State. 4 and 8)
10. $\triangle ABC \sim \triangle DEF$	10. AA ~ AA

Theorem
4.5

(SSS ~ SSS) If corresponding sides of two triangles are pro-portional, then the two triangles are similar.

Given: $\dfrac{AC}{DF} = \dfrac{AB}{DE} = \dfrac{BC}{EF}$

Prove: $\triangle ABC \sim \triangle DEF$

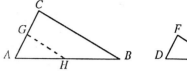

Note: Without loss of generality, we assume that $DF < AC$ (see the note in the proof of Theorem 4.4).

Proof:

Statements	Reasons
1. Let G be the point of $\overline{AC}$ such that $AG = DF$. Let H be the point of $\overline{AB}$ such that $AH = DE$.	1. Ruler Post.
2. $\dfrac{AC}{DF} = \dfrac{AB}{DE} = \dfrac{BC}{EF}$	2. Given
3. $\dfrac{AC}{AG} = \dfrac{AB}{AH} = \dfrac{BC}{EF}$	3. Substitution
4. $\angle A \simeq \angle A$	4. Reflexivity
5. $\triangle AHG \sim \triangle ABC$	5. SAS ~ SAS
6. $\dfrac{BC}{HG} = \dfrac{AB}{AH}$	6. Def. of ~
7. $\dfrac{BC}{HG} = \dfrac{BC}{EF}$	7. Transitivity
8. $EF = HG$	8. Mult. Axiom of =
9. $\triangle AHG \simeq \triangle DEF$	9. SSS = SSS
10. $\angle F \simeq \angle AGH,\ \angle E \simeq \angle AHG$	10. Corr. parts of $\simeq \triangle$s are $\simeq$
11. $\angle AGH \simeq \angle C,\ \angle AHG \simeq \angle B$	11. Def. of ~
12. $\angle F \simeq \angle C,\ \angle E \simeq \angle B$	12. Transitivity
13. $\triangle ABC \sim \triangle DEF$	13. AA ~ AA

In the sections in which congruent triangles are discussed, we described the following relationships that determine the congruence of triangles.

$$\text{SAS} = \text{SAS} \quad \text{(Postulate 2.1)}$$
$$\text{ASA} = \text{ASA} \quad \text{(Postulate 2.2)}$$
$$\text{SSS} = \text{SSS} \quad \text{(Postulate 2.3)}$$
$$\text{SAA} = \text{SAA} \quad \text{(Exercise Set 3.3, No. 48)}$$

In this chapter, we discuss the following relationships that determine the similarity of triangles.

$$\text{AAA} \sim \text{AAA} \quad \text{(Postulate 4.1)}$$
$$\text{AA} \sim \text{AA} \quad \text{(Theorem 4.1)}$$
$$\text{SAS} \sim \text{SAS} \quad \text{(Theorem 4.4)}$$
$$\text{SSS} \sim \text{SSS} \quad \text{(Theorem 4.5)}$$

Differentiating between congruence relationships and similarity relationships is important: If two triangles are congruent, they are also similar; but if two triangles are similar, they are not necessarily congruent. Examples 1 and 2 illustrate the use of Theorems 4.4 and 4.5.

Example 1

Given: $\overline{AE}$ and $\overline{BD}$ bisect each other

Prove: $\triangle ABC \sim \triangle EDC$

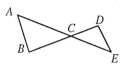

Solution

Proof:

Statements	Reasons
1. $\overline{AE}$ and $\overline{BD}$ bisect each other	1. Given
2. $AC = CE$, $BC = CD$	2. Def. of bisector
3. $\dfrac{AC}{CE} = 1$, $\dfrac{BC}{CD} = 1$	3. Mult. Axiom of =
4. $\dfrac{AC}{CE} = \dfrac{BC}{CD}$	4. Substitution
5. $\angle ACB \cong \angle ECD$	5. Vert. $\angle$s are $\cong$
6. $\triangle ABC \sim \triangle EDC$	6. SAS $\sim$ SAS

Example 2

Given: $\triangle ABC$ and $\triangle BCD$ as shown in Figure 4.5

Prove: $\triangle ABC \sim \triangle BDC$.

Figure 4.5

Solution

Notice that $\dfrac{AC}{BC} = \dfrac{BC}{DC} = \dfrac{AB}{BD} = \dfrac{2}{1}$. Therefore, since SSS $\sim$ SSS, $\triangle ABC \sim \triangle BCD$.

EXERCISE SET 4.3

In Exercises 1–6, determine whether the two triangles are similar. If so, state why they are similar and write the similarity.

1.

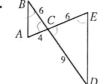

2.

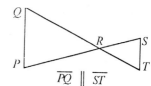

$\overline{PQ} \parallel \overline{ST}$

3.

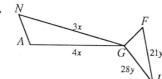

4.

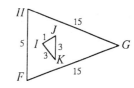

5.

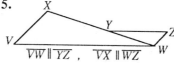

$\overline{VW} \parallel \overline{YZ}$, $\overline{VX} \parallel \overline{WZ}$

6.

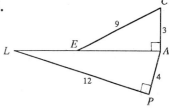

Figures for Exercises 7–10

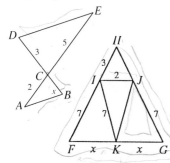

In Exercises 7–10, find x.

7. $\triangle ABC \sim \triangle EDC$

8. $\triangle ABC \sim \triangle DEC$

9. $\triangle FGH \sim \triangle IJK$

10. $\triangle FGH \sim \triangle GJK$

Figure for Exercise 11

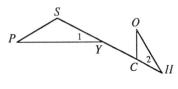

11. If $S-Y-C-H$, $SC = 10$, $HY = 8$, $PY = 7$, $HO = 4$, and if $\triangle PSY \sim \triangle OCH$, find CY.

Figure for Exercise 12

12. If $A-E-B$, $D-F-C$, $AB = 12$, $CD = 23$, $BC = 5$, $AD = 19$, $DF = 2 \cdot BE$, and if $\triangle ADE \sim \triangle CBF$, find BE.

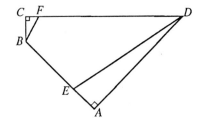

Prove the following.

13. If two triangles are similar, the lengths of the corresponding altitudes have the same ratio as the lengths of the corresponding sides.

14. Repeat Exercise 13 for corresponding medians.

15. Repeat Exercise 13 for corresponding angle bisectors.

16. Two triangles are similar if their corresponding sides are respectively parallel to each other.

17. Two triangles are similar if their corresponding sides are respectively perpendicular to each other.

18. If three or more parallel lines are cut by two transversals, the corresponding segments of the transversals are proportional. (Hint: There are two cases: (a) the transversals are parallel, (b) the transversals are not parallel.)

19. A bisector of an angle of a triangle divides the opposite side into segments that are proportional to the adjacent sides. (Hint: Given $\triangle ABC$ with $\overrightarrow{AD}$ bisecting $\angle A$ and $B-D-C$, let k be a line parallel to $\overleftrightarrow{AD}$ through C and let $\overleftrightarrow{BA} \cap k = \{E\}$. Notice that $BD/DC = BA/AE$ and that $\angle AEC \simeq \angle ACE$.)

4.4 REGULAR POLYGONS

In Chapter 3, we defined a regular polygon as a polygon in which all the angles and all the sides are congruent. Other parts of a regular polygon are its *center, central angle, radius,* and *apothem.*

Definition

> The **center** of a regular polygon is the unique point equidistant from each of the vertices of the polygon.

The center of a regular polygon is the center of a circle containing all the vertices of the polygon.

Definition	A **central angle** of a regular polygon is an angle with vertex at the center of the polygon and containing two adjacent vertices of the polygon.

Definition	A **radius** of a regular polygon is a line segment with endpoints at the center of the polygon and a vertex of the polygon.

Definition	An **apothem** of a regular polygon is a line segment perpendicular to one of the sides with one endpoint on the side and the other endpoint at the center of the polygon.

Example 1

Figure 4.6

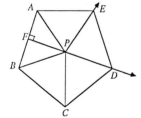

In Figure 4.6, suppose that *ABCDE* is a regular pentagon and that *AP* = *BP* = *CP* = *DP* = *EP*. Name the center, the central angles, the radii, and the apothem shown.

Solution

The center is *P*; the central angles are ∠*APE*, ∠*DPE*, ∠*CPD*, ∠*BPC*, ∠*APB*; the radii are $\overline{AP}$, $\overline{BP}$, $\overline{CP}$, $\overline{DP}$, $\overline{EP}$; and the apothem shown is $\overline{FP}$.

Example 2

Find the measure of a central angle of a regular polygon with 20 sides.

Solution

Since the central angles of a regular polygon are congruent, the measure of a central angle is $360°/20 = 18°$.

Example 3

If the measure of a central angle of a regular polygon is 36°, find the number of sides in the polygon.

Solution

The number of sides is 360°/36° = 10.

■

Example 4

Find the perimeter of a regular decagon if each side is 14 meters long.

Solution

The perimeter is 10(14 meters) = 140 meters.

■

Regular polygons have many special properties not shared by polygons in general. Some of these properties are illustrated below and others are left as exercises.

Example 5

Given: Regular hexagon *ABCDEF* with center *P*

Prove: $\overrightarrow{EP}$ bisects ∠*DEF*

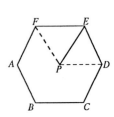

Solution

Proof:

Statements	Reasons
1. Regular hexagon *ABCDEF* with center *P*	1. Given
2. *FP* = *DP*	2. Def. of center
3. *EP* = *EP*	3. Reflexivity

4. $FE = DE$	**4.** Def. of regular polygon
5. $\triangle FEP \simeq \triangle DEP$	**5.** SSS = SSS
6. $\angle FEP \simeq \angle DEP$	**6.** Corr. parts of $\simeq \triangle$s are $\simeq$
7. $\overrightarrow{EP}$ bisects $\angle DEF$	**7.** Def. of angle bisector

■

Example 6

Given: Regular pentagon $ABCDE$ with center P,
$\overline{PF} \perp \overline{AB}$

Prove: $\overline{PF}$ bisects $\overline{AB}$

Solution

Proof:

Statements	Reasons
1. Regular pentagon $ABCDE$ with center P, $\overline{PF} \perp \overline{AB}$	**1.** Given
2. $AP = BP$	**2.** Def. of center
3. $m\angle PAB = m\angle PBA$	**3.** If two sides of a $\triangle$ are $\simeq$, the $\angle$s opp. these sides are $\simeq$
4. $m\angle AFP = m\angle BFP$	**4.** Def. of $\perp$
5. $\triangle AFP \simeq \triangle BFP$	**5.** SAA = SAA
6. $AF = BF$	**6.** Corr. parts of $\simeq \triangle$s are $\simeq$
7. $\overline{PF}$ bisects $\overline{AB}$	**7.** Def. of segment bisector

■

Thus, we can prove that a radius of a regular polygon serves essentially as an angle bisector of an interior angle and an apothem of a regular polygon bisects the side it intersects.

The following constructions illustrate regular polygons with three or four sides.

Construction 4.3	Construct an equilateral triangle.

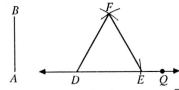

Given: Arbitrary $\overline{AB}$

Construct: $\triangle DEF$ so that
$$DE = EF = FD = AB$$

Step 1. Draw a line $\overleftrightarrow{DQ}$, and use Construction 1.1 to find a segment $\overline{DE}$ on $\overleftrightarrow{DQ}$ so that $DE = AB$.

Step 2. Using AB as the radius, draw circle D and circle E. Label one of the points of intersection of the circles F.

Step 3. Draw $\overline{DF}$ and $\overline{EF}$. Then $\triangle DEF$ is the required triangle.

Construction 4.4	Construct a square.

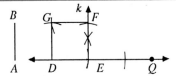

Given: Arbitrary $\overline{AB}$

Construct: Square $DEFG$

Step 1. Draw a line $\overleftrightarrow{DQ}$, and use Construction 1.1 to find a segment $\overline{DE}$ on $\overleftrightarrow{DQ}$ so that $DE = AB$.

Step 2. Use Construction 1.5 to find the line k perpendicular to $\overleftrightarrow{DQ}$ at E.

Step 3. Use Construction 1.1 to find a point F so that $EF = AB$.

Step 4. Using AB as the radius, draw circle D and circle F. One of the points of intersection will be E. Label the other point G.

Step 5. Draw $\overline{DG}$ and $\overline{FG}$. Thus, $DEFG$ is the required square.

EXERCISE SET 4.4

In Exercises 1–3, draw a regular hexagon $ABCDEF$ with center P.

1. Name all the radii.

2. Name all the central angles.

3. Draw all the apothems.

Figure for Exercises 4–7

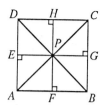

In Exercises 4–7, square *ABCD* is given.

4. Name all the diagonals.

5. Name all the radii.

6. Name all the central angles.

7. Name all the apothems.

In Exercises 8–16, find the measure of a central angle of each regular polygon with the given number of sides. (See Exercise 39.)

8. 3	11. 6	14. 9
9. 4	12. 7	15. 10
10. 5	13. 8	16. n

In Exercises 17–25, if the measure of a central angle of a regular polygon is given, find the number of sides in the polygon.

17. 60°	20. 10°	23. 21° 10′ 35 5/17″
18. 30°	21. 5°	24. $(360/n)°$
19. 24°	22. 1°	25. $[5/(2n)]°$

In Exercises 26–32, construct regular polygons with the given number of sides. (Hint: Exercises 28–32 may be constructed by bisecting central angles of Exercise 26 or Exercise 27.)

26. 3	29. 8	32. 24
27. 4	30. 12	
28. 6	31. 16	

33. Construct a square with a given side of 1.5 inches.

34. Construct an equilateral triangle with a given side of 1.75 inches.

35. Construct a square with a side congruent to the altitude of an equilateral triangle with a given side of 2 inches.

36. Construct an equilateral triangle with a side congruent to the diagonal of a square with a given side of 1 inch.

In Exercises 37–42, find the perimeters of the regular polygons if each of the sides has the given length.

37. Triangle, 17

39. Hexagon, 3

41. Octagon, 29.2

38. Pentagon, 4

40. Heptagon, $\sqrt{3}$

42. n-gon, 67.35

Prove the following.

43. Central angles of a regular polygon are congruent.

44. Apothems of a regular polygon are congruent.

45. Central angles of two regular pentagons are congruent.

Chapter 4
Key Terms

Chapter 4
Review Exercises

Fill each blank with the word *always, sometimes,* or *never.*

1. A ratio is _____ a proportion.

2. Similar triangles are _____ congruent.

3. Congruent triangles are _____ similar.

4. A line parallel to one side of a triangle _____ intersects the other two sides in distinct points.

5. A line that intersects two sides of a triangle in distinct points and divides these two sides into proportional segments is _____ parallel to the third side.

6. If two triangles are similar, the two triangles are _____ congruent.

7. The center of a regular polygon is _____ in the exterior of the polygon.

8. Two apothems of a regular polygon are _____ perpendicular.

9. Two apothems of a regular polygon are _____ parallel.

10. An angle with vertex at the center of a regular polygon and containing two vertices of the polygon is _____ a central angle.

True-False: If the statement is true, mark it so. If it is false, replace the underlined word so as to make a true statement.

11. A <u>ratio</u> is a statement that two <u>proportions</u> are equal.

12. In the proportion $a/b = b/c$, b is the <u>second proportional</u> between a and c.

13. Two polygons are <u>similar</u> if corresponding angles are congruent and the corresponding sides are proportional segments.

14. If the three angles of one triangle are congruent to the three angles of another triangle, then the triangles are <u>congruent</u>.

15. If corresponding angles of two triangles are <u>proportional</u>, then the two triangles are similar.

16. An apothem of a regular polygon is a line segment with endpoints at the center of the polygon and a vertex of the polygon.

17. The vertex of a regular polygon is the unique point which is equidistant from each of the centers of the polygon.

18. The perpendicular bisector of a side of a regular polygon contains the center of the polygon.

19. The bisector of an angle of a regular polygon must contain two vertices of the polygon.

20. A central angle of a regular octagon has twice the measure of a central angle of a regular quadrilateral.

Answer the following questions.

In Exercises 21–24, express each ratio as a fraction and simplify.

21. 6:4

22. 1/2:2

23. The ratio of 5 yards to 2 miles

24. The ratio of 5 gallons to 4 quarts

In Exercises 25–28, solve for x in each of the proportions.

25. $x/3 = 5/7$

26. $2/x = 9/4$

27. $5/(x + 1) = 9/(x + 2)$

28. $6:7x = 5:3x$

In Exercises 29–30, find the fourth proportional.

29. 2, 3, 4

30. $7x, 10x, 17x$

In Exercises 31–32, find the mean proportional between:

31. 1 and 169

32. 2/7 and 5/7

In Exercises 33–34, write the congruences between the corresponding angles and the continued proportion for the corresponding sides.

33. $\triangle ACB \sim \triangle DEF$ 34. $PQRST \sim JKLMN$

35. Find the height of a tower that casts a 300-foot shadow if an adjacent building 800 feet tall casts a 500-foot shadow.

In Exercises 36–39, given $\triangle KMN$, $\overline{PL} \parallel \overline{NM}$, $K-P-N$, $K-L-M$.

Figure for Exercises 36–39

36. Find LM if $KP = 5$, $KL = 7$, and $PN = 3$.

37. Find PN if $KP = 4$, $KM = 12$, and $KL = 5$.

38. Find KL if $KM = 9$ and $KP = 2PN$.

39. Find PL if $NM = 4$ and $KN = 2KP$.

In Exercises 40–42, given $\triangle ABC$, $\triangle DEF$, $\overline{DE} \parallel \overline{AC}$, $\overline{EF} \parallel \overline{BC}$.

40. Find $m\angle D$ if $\angle F \cong \angle B$.

41. Find $m\angle F$ if $\angle F \cong \angle B$.

42. Find DF if $DE/AC = EF/CB = 2/3$.

In Exercises 43–44, given $\triangle ABC$, $A-M-C$, $B-N-C$, $\overline{MN} \parallel \overline{AB}$, $AC = BC = 10$, $AM = 4$, $MC = AB$.

43. Find MN.

44. Find $m\angle AMB - m\angle MBN$ if $m\angle C = 27°$.

In Exercises 45–46, given $\triangle ABC$, $\triangle RST$, $m\angle A = m\angle T$, $m\angle C = m\angle S$.

45. Find AB if $BC/RS = 2/3$ and $RT = 9$.

46. What must the ratio of BC/RS equal in order that $\triangle RST \cong \triangle BCA$?

47. If $ABCDEF$ is a regular hexagon, and $AD = 12$, find BC.

48. The measure of a central angle of one regular polygon is 4/5 of the measure of a central angle of a second regular polygon. If the first polygon has three more sides than the second, how many sides does it have?

49. An equilateral triangle and a regular octagon have sides of equal length. What is the ratio of their perimeters?

Proofs:

In Exercises 50–51, given $\triangle ABC$, $D-E-F$, and $\square ACFD$.

50. *Prove:* $\triangle BDE \sim \triangle CFE$

51. *Prove:* $CF/DB = CE/EB$

In Exercises 52–53, given $\triangle ABC$, $D-E-F$, $m\angle B = m\angle FCE$, $AD = 3DB$, and $CE = 3EB$.

52. *Prove:* $\triangle DBE \sim \triangle FCE$

53. *Prove:* $\triangle FCE \sim \triangle ABC$

54. *Prove:* The bisector of an interior angle of a triangle divides the opposite side into two segments that are proportional to the adjacent sides.

   Given: $\triangle ABC$, $\overrightarrow{CD}$ bisects $\angle ACB$

   Prove: $AD/DB = AC/CB$

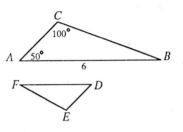

Figure for Exercises 40–42

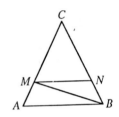

Figure for Exercises 43–44

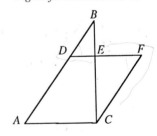

Figure for Exercises 50–53

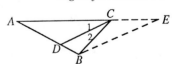

Figure for Exercise 54

55. Given: $\triangle PQR$, $P-R-S$,
    $\overrightarrow{RT}$ bisects $\angle QRS$

    Prove: $PT/QT = PR/QR$

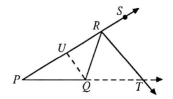

56. Prove that any two squares are similar.

Constructions:

57. Construct a hexagon similar to a given nonregular hexagon.

58. Construct angles whose measures are 60°, 30°, and 15°.

59. Construct a square with a given side of 2 inches.

60. Construct a regular hexagon with a given side of 1.5 inches.

Right Triangles

Historical Note

There are well over 300 known proofs of the Pythagorean Theorem. Pythagoras was a pupil of Thales, who is believed to have developed the first proof of this theorem. Through the years, many other people, including President James A. Garfield, have developed a wide variety of proofs of the same theorem.

Courtesy of Historical Pictures Service, Chicago

Thales (640?–546 B.C.)

Pythagorean Theorem

The square of the length of the hypotenuse of a right triangle equals the sum of the squares of the lengths of the legs.

Having discussed the concept of congruent triangles in Chapter 2, we now address the special case of right triangles. Because this concept requires knowledge of radicals and quadratic equations from algebra, an optional review section is provided, as well as an optional section (Section 5.5) covering basic trigonometric concepts.

5.1   REVIEW OF RADICALS AND QUADRATIC EQUATIONS (Optional)

In succeeding chapters, we use quadratic equations and expressions involving radicals. Section 5.1 should prove useful to the reader who is out of practice with these types of algebraic manipulations and may be omitted by the reader who feels amply qualified to do the work without additional practice.

In this course, we are concerned primarily with square roots, so we restrict our attention to square root radicals.

Given a real number $r \geq 0$, the **square root of r** (written $\sqrt{r}$) is the positive solution to the equation $x^2 - r = 0$. The **negative square root of r** (written $-\sqrt{r}$) is the negative solution to the equation $x^2 - r = 0$. Thus $\sqrt{9} = 3$ and $-\sqrt{9} = -3$ are solutions of the equation $x^2 - 9 = 0$. For many choices of r, the equation $x^2 - r = 0$ will not have a rational solution. In this event, we may approximate the value of $\sqrt{r}$ through the use of a calculator or table. (See Appendix G.) In this book, we are usually more interested in exact answers than in approximate answers.

Example 1

Find the exact solutions and the approximate solutions to the equation $x^2 - 3 = 0$.

Solution

The exact solutions are $\sqrt{3}$ and $-\sqrt{3}$.
The approximate solutions are 1.732 and −1.732.

Example 2

Simplify $\sqrt{12}$ and $\sqrt{612}$.

Solution

$$\sqrt{12} = \sqrt{4 \cdot 3} = \sqrt{4} \cdot \sqrt{3} = 2\sqrt{3}$$
$$\sqrt{612} = \sqrt{36 \cdot 17} = \sqrt{36} \cdot \sqrt{17} = 6\sqrt{17}$$

■

Each simplification is obtained by factoring the original number into the largest **perfect square** factor of the number (i.e., $4 = 2^2$, $36 = 6^2$) times the remaining factor of the number that is not a perfect square (i.e., 3, 17).

■

Example 3

Simplify $\dfrac{1}{\sqrt{3}}$ and $\sqrt{\dfrac{5}{7}}$.

Solution

$$\frac{1}{\sqrt{3}} = \frac{1}{\sqrt{3}} \cdot \frac{\sqrt{3}}{\sqrt{3}} = \frac{\sqrt{3}}{3}.$$

$$\sqrt{\frac{5}{7}} = \sqrt{\frac{5}{7} \cdot \frac{7}{7}} = \sqrt{\frac{35}{49}} = \frac{\sqrt{35}}{\sqrt{49}} = \frac{\sqrt{35}}{7}.$$

■

We prefer the expression $\dfrac{\sqrt{3}}{3}$ to the expression $\dfrac{1}{\sqrt{3}}$, since for computational purposes, $\dfrac{1}{1.732}$ is harder to evaluate than $\dfrac{1.732}{3}$. Similarly, we prefer $\dfrac{\sqrt{35}}{7}$ to $\sqrt{\dfrac{5}{7}}$.

A **quadratic equation** is an equation of the form $ax^2 + bx + c = 0$, where a, b, and c are real numbers, with $a \neq 0$. The following theorem from algebra is useful in solving quadratic equations.

Theorem A

> If m and n are real numbers and $mn = 0$, then $m = 0$ or $n = 0$.

Proof:

If $m = 0$, we are through. If $m \neq 0$, use the multiplication axiom of equality to multiply both sides of $mn = 0$ by $1/m$. The result is $n = 0$.

We use Theorem A in the following example.

Example 4

Solve for x: $5x^2 + 7x - 6 = 0$.

Solution

Factor the left side and get $(5x - 3)(x + 2) = 0$.
Thus, by Theorem A: $5x - 3 = 0$ or $x + 2 = 0$.
By the Addition Axiom of Equality: $5x = 3$ or $x = -2$.
By the Multiplication Axiom of Equality: $x = 3/5$ or $x = -2$.
Therefore, the solution set is $\{-2, 3/5\}$.

■

A simple form of a quadratic equation that often occurs is $x^2 = k^2$. We solve this type of quadratic equation using the following theorem.

Theorem B

If $x^2 = k^2$, then $x = \pm k$.

Proof:

Statements	Reasons
1. $x^2 = k^2$	1. Given
2. $x^2 - k^2 = 0$	2. Add. Axiom of $=$
3. $(x + k)(x - k) = 0$	3. Substitution
4. $x + k = 0$ or $x - k = 0$	4. Theorem A
5. $x = \pm k$	5. Add. Axiom of $=$

Example 5

Solve for x: $x^2 = 121$.

Solution

The solution set is $\{\pm 11\}$.

■

Example 6

Solve for y: $y^2 = 13$.

Solution

The solution set is $\{\pm\sqrt{13}\}$.

◼

If the solutions of a quadratic equation are not rational, then the equation is not factorable. In this event, we use the following theorem.

Theorem
C

$$\text{If } ax^2 + bx + c = 0 \text{ and } a \neq 0, \text{ then } x = \frac{-b \pm \sqrt{b^2 - 4ac}}{2a}.$$

Proof:

Statements	Reasons
1. $ax^2 + bx + c = 0, a \neq 0$	1. Given
2. $x^2 + \dfrac{b}{a}x + \dfrac{c}{a} = 0$	2. Mult. Axiom of $=$
3. $x^2 + \dfrac{b}{a}x = -\dfrac{c}{a}$	3. Add. Axiom of $=$
4. $x^2 + \dfrac{b}{a}x + \dfrac{b^2}{4a^2} = \dfrac{b^2}{4a^2} - \dfrac{c}{a}$	4. Add. Axiom of $=$
5. $\left(x + \dfrac{b}{2a}\right)^2 = \dfrac{b^2 - 4ac}{4a^2}$	5. Substitution
6. $x + \dfrac{b}{2a} = \dfrac{\pm\sqrt{b^2 - 4ac}}{2a}$	6. Theorem B
7. $x = \dfrac{-b \pm \sqrt{b^2 - 4ac}}{2a}$	7. Add. Axiom of $=$

The expression for x in Theorem C is the **quadratic formula**. The procedure used in the proof is known as **completing the square**. The term $b^2/(4a^2)$, which was added to both sides of the equation in Step 4, was obtained by taking half of the coefficient of x and squaring the result, namely $b^2/(4a^2) = [b/(2a)]^2$. Thus, the left side of the equation became a perfect square, as shown in Step 5.

Example 7

Solve for x: $3x^2 - 4x - 5 = 0$.

Solution

By Theorem C: $x = \dfrac{4 \pm \sqrt{16 - 4(-15)}}{6}$, which simplifies to

$x = \dfrac{4 \pm \sqrt{76}}{6} = \dfrac{4 \pm \sqrt{4 \cdot 19}}{6} = \dfrac{4 \pm 2\sqrt{19}}{6} = \dfrac{2 \pm \sqrt{19}}{3}$.

EXERCISE SET **5.1**

In Exercises 1–9, which numbers are *perfect squares*?

1. 4	4. 121	7. 324
2. 20	5. 999	8. 5929
3. 49	6. 289	9. 1481089

In Exercises 10–18, simplify the radicals.

10. $\sqrt{18}$	13. $\sqrt{867}$	16. $\sqrt{7744}$
11. $\sqrt{92}$	14. $\sqrt{2527}$	17. $\sqrt{23625}$
12. $\sqrt{1225}$	15. $\sqrt{899}$	18. $\sqrt{47753}$

In Exercises 19–27, simplify the radical expressions.

19. $1/\sqrt{7}$	22. $\sqrt{6/9}$	25. $2\sqrt{289/53}$
20. $1/\sqrt{81}$	23. $\sqrt{3/2}$	26. $(7\sqrt{722/1250})/2$
21. $1/\sqrt{13}$	24. $\sqrt{9/13}$	27. $(13\sqrt{191/270})/5$

In Exercises 28–36, solve for x:

28. $x^2 = 25$	31. $x^2 + 4x + 4 = 0$	34. $72 - x^2 = 6x$
29. $x^2 = 7$	32. $x^2 - 4x + 4 = 0$	35. $2x^2 + 3x - 8 = 0$
30. $x^2 - 2 = 0$	33. $x^2 - 4x = 21$	36. $5x^2 - 7x = 64$

In Exercises 37–45, solve the proportions for y.

37. $\dfrac{3}{y} = \dfrac{y}{12}$	38. $\dfrac{y}{5} = \dfrac{7}{y}$	39. $\dfrac{9}{y} = \dfrac{y}{13}$

$$40. \ \frac{y+2}{2} = \frac{2}{y} \qquad 42. \ \frac{2}{y} = \frac{2y-1}{4} \qquad 44. \ \frac{3y-7}{4} = \frac{4}{y+3}$$

$$41. \ \frac{y-3}{6} = \frac{6}{y} \qquad 43. \ \frac{2y+3}{9} = \frac{9}{y} \qquad 45. \ \frac{4-y}{7} = \frac{5}{3-2y}$$

5.2 RIGHT TRIANGLE CONGRUENCE THEOREMS

In Section 2.6, we defined *right triangle, hypotenuse,* and *leg.* In this section, we discuss congruence theorems that refer only to right triangles. If we let *L* represent *leg, H* represent *hypotenuse,* and *A* represent *acute angle,* then the following are true (the reader should translate the statements into complete sentences):

1. LA = LA

2. HA = HA

3. LL = LL

Notice that (1) is a special case of ASA = ASA or SAA = SAA; (2) is a special case of SAA = SAA; and (3) is a special case of SAS = SAS.

The following theorem has no counterpart in the section on congruent triangles (i.e., there is no SSA = SSA theorem).

Theorem 5.1

(**HL = HL**) If the hypotenuse and a leg of one right triangle are congruent to the corresponding parts of another right triangle, then the two right triangles are congruent.

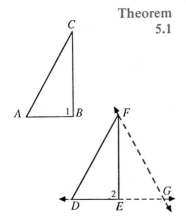

Given: ∠1 and ∠2 are right ∠s,
  $AC = DF, BC = EF$

Prove: △$ABC ≃$ △DEF

Proof:

Statements	Reasons
1. ∠1 and ∠2 are right ∠s, $AC = DF, BC = EF$	1. Given
2. Let G be a point on $\overleftrightarrow{DE}$ such that $D-E-G$ and $GE = AB$	2. Ruler Post. and def. of between

3. There is exactly one line that contains *F* and *G*.	3. Any 2 distinct points in space have exactly one line that contains them.
4. ∠*FEG* is a right ∠	4. Def. of meas. and Angle Add. Post.
5. △*ABC* ≃ △*GEF*	5. LL = LL
6. *AC* = *GF*	6. Corr. parts of ≃ △s are ≃.
7. *DF* = *GF*	7. Substitution (State. 1 and 6)
8. ∠*EDF* ≃ ∠*EGF*	8. If 2 sides of a △ are ≃, the ∠s opp. these sides are ≃.
9. △*GEF* ≃ △*DEF*	9. HA = HA
10. △*ABC* ≃ △*DEF*	10. Transitivity

In the last step, we assume that transitivity holds for congruence of triangles (i.e., if △*ABC* ≃ △*GEF* and △*GEF* ≃ △*DEF*, then △*ABC* ≃ △*DEF*). The proof is left as an exercise.

Example

Given: $\overline{BD} \perp \overline{AC}, \overline{AE} \perp \overline{BC}$
$AD = BE$

Prove: △*ABC* is isosceles

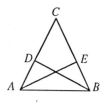

Solution

Proof:

Statements	Reasons
1. $\overline{BD} \perp \overline{AC}, \overline{AE} \perp \overline{BC}, AD = BE$	1. Given
2. *AB* = *AB*	2. Reflexivity
3. △*ABD* ≃ △*BAE*	3. HL = HL
4. ∠*DAB* ≃ ∠*EBA*	4. Corr. parts of ≃ △s are ≃.
5. *AC* = *BC*	5. If 2 ∠s of a △ are ≃, the sides opp. the ∠s are ≃.
6. △*ABC* is isosceles	6. Def. of isosceles

EXERCISE SET 5.2

In Exercises 1–3, state and prove the theorems.

1. LA = LA   2. HA = HA   3. LL = LL

4. Prove that two right triangles are similar if an acute angle of one has the same measure as an acute angle of the other.

Figure for Exercises 5–6

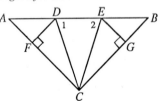

5. *Given:* $AD = BC$, $\overline{AD} \perp \overline{AB}$, $\overline{CD} \perp \overline{BC}$
   Prove: $\triangle ABD \simeq \triangle CDB$

6. *Given:* $\overline{AB} \parallel \overline{CD}$, $AB = CD$, $\overline{BC} \perp \overline{CD}$
   Prove: $\triangle ABD$ is a right $\triangle$

7. *Given:* $\overline{DE}$ bisects $\overline{AC}$ at B, $\overline{DE} \perp \overline{AE}$, $\overline{DE} \perp \overline{DC}$
   Prove: $AE = DC$

8. *Given:* $\overline{DE} \perp \overline{AE}$, $\overline{AE} \parallel \overline{DC}$, $DB = EB$, $A–B–C$
   Prove: $BC = BA$

9. *Given:* $\overline{DE} \perp \overline{AE}$, $\overline{DE} \perp \overline{DC}$, $AE = CD$, $BD = BE$
   Prove: $m\angle A = m\angle C$

10. *Given:* $\overline{DF} \perp \overline{AC}$, $\overline{EG} \perp \overline{BC}$, $DF = EG$, $\angle 1 \simeq \angle 2$
    Prove: $\angle ACD \simeq \angle BCE$

11. *Given:* $\overline{DF} \perp \overline{AC}$, $\overline{EG} \perp \overline{BC}$, $\angle BAC \simeq \angle ABC$, $CF = CG$
    Prove: $\angle 1 \simeq \angle 2$

In Exercises 12–14, prove that congruence of triangles is

12. Reflexive

13. Symmetric

14. Transitive

5.3 SOME PROPERTIES OF RIGHT TRIANGLES

Right triangles have many interesting properties. In this section, we discuss theorems that describe medians and altitudes extending from the right angle of a right triangle.

Theorem 5.2	The median to the hypotenuse of any right triangle is half as long as the hypotenuse.

Given: $\overline{AC} \perp \overline{BC}$, $DA = BD$

Prove: $CD = BA/2$

Proof:

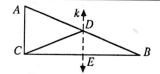

Statements	Reasons
1. $\overline{AC} \perp \overline{BC}$, $DA = BD$	1. Given
2. There is exactly 1 line k through D such that $k \parallel \overleftrightarrow{AC}$. Let $k \cap \overleftrightarrow{BC} = \{E\}$.	2. Parallel Post.
3. $CE = BE$	3. A line $\parallel$ to 1 side of a $\triangle$ that intersects the other 2 sides in distinct points divides these 2 sides into prop. segs.
4. $\overline{DE} \perp \overline{BC}$	4. If a line in a plane is $\perp$ to 1 of 2 $\parallel$ lines in the same plane, then it is also $\perp$ to the other
5. $\angle CED \simeq \angle BED$	5. Def. of $\perp$
6. $DE = DE$	6. Reflexivity
7. $\triangle CED \simeq \triangle BED$	7. SAS = SAS
8. $BD = CD$	8. Corr. parts of $\simeq \triangle$s are $\simeq$
9. $BD + DA = BA$	9. Def. of between
10. $2BD = BA$	10. Substitution (State. 1 and 9)
11. $2CD = BA$	11. Substitution (State. 8 and 10)
12. $CD = BA/2$	12. Mult. Axiom of =

Example 1

Find the perimeter of $\triangle ABC$ in Figure 5.1, given that $\overline{DE}$ and $\overline{DF}$ are medians.

Figure 5.1

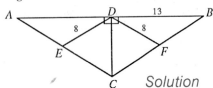

Solution

By Theorem 5.2, since *DE* and *DF* are half of *AC* and *BC*, respectively, it follows that $AC = BC = 16$. Since $CD = CD$ by reflexivity, $\triangle ACD \simeq \triangle BCD$ by HL = HL. Hence, $AD = BD = 13$. Therefore, the perimeter of $\triangle ABC = 16 + 16 + 13 + 13 = 58$. ∎

Theorem 5.3	In any right triangle, the altitude to the hypotenuse forms two right triangles that are similar to each other and to the original triangle.

Given: $\overline{AC} \perp \overline{BC}$, $\overline{CD} \perp \overline{AB}$

Prove: $\triangle ACD \sim \triangle CBD$,
$\triangle ACD \sim \triangle ABC$,
$\triangle CBD \sim \triangle ABC$

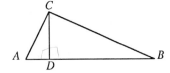

Proof:

Statements	Reasons
1. $\overline{AC} \perp \overline{BC}$, $\overline{CD} \perp \overline{AB}$	1. Given
2. $\angle A \simeq \angle A$	2. Reflexivity
3. $\angle ADC \simeq \angle ACB$	3. Def. of $\perp$ and meas.
4. $\triangle ACD \sim \triangle ABC$	4. AA ~ AA
5. $\angle B \simeq \angle B$	5. Reflexivity
6. $\angle BDC \simeq \angle BCA$	6. Def. of $\perp$ and meas.
7. $\triangle CBD \sim \triangle ABC$	7. AA ~ AA
8. $\triangle ABC \sim \triangle CBD$	8. Symmetry
9. $\triangle ACD \sim \triangle CBD$	9. Transitivity (State. 4 and 8)

Theorem 5.4 is a corollary of Theorem 5.3, but we prefer to treat 5.4 as a theorem to emphasize its importance. This corollary/theorem is useful in many computational situations that result in quadratic equations.

Theorem 5.4	Given a right triangle, the altitude to the hypotenuse divides the hypotenuse into two segments such that (1) the altitude is the geometric mean of these segments, (2) each leg is the geometric mean of the hypotenuse and the segment of the hypotenuse adjacent to the leg.

Given: $\overline{AC} \perp \overline{BC}$, $\overline{AB} \perp \overline{CD}$

Prove: $\dfrac{AD}{CD} = \dfrac{CD}{BD}$, $\dfrac{AB}{AC} = \dfrac{AC}{AD}$, $\dfrac{AB}{BC} = \dfrac{BC}{BD}$

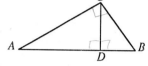

Proof:

Statements	**Reasons**
1. $\overline{AC} \perp \overline{BC}$, $\overline{AB} \perp \overline{CD}$	1. Given
2. $\triangle ADC \sim \triangle CDB$	2. In any right $\triangle$, the alt. to the hyp. forms 2 right $\triangle$s that are $\sim$ to each other and to the original $\triangle$
3. $\dfrac{AD}{CD} = \dfrac{CD}{BD}$	3. Def. of $\sim$
4. $\triangle ACB \sim \triangle ADC$	4. Same as Reason 2
5. $\dfrac{AB}{AC} = \dfrac{AC}{AD}$	5. Def. of $\sim$
6. $\triangle ACB \sim \triangle CDB$	6. Same as Reason 2
7. $\dfrac{AB}{BC} = \dfrac{BC}{BD}$	7. Def. of $\sim$

Example 2

Figure 5.2

In Figure 5.2, find AD if $\overline{BD}$ is an altitude of $\triangle ABC$, $m\angle ABC = 90°$, $AB = 10$, and $CD = 5$.

Solution

By Theorem 5.4, $\dfrac{AC}{AB} = \dfrac{AB}{AD}$. Since $AC = AD + DC$, we have $\dfrac{AD + 5}{10} = \dfrac{10}{AD}$.
Hence, $(AD)^2 + 5(AD) = 100$, and so $(AD)^2 + 5(AD) - 100 = 0$. Using the quadratic formula, we find that $AD = \dfrac{-5 \pm \sqrt{25 + 400}}{2}$. Simplifying, we

get $AD = \dfrac{-5 \pm 5\sqrt{17}}{2}$. Since the length of a line segment is never

negative, $AD = \dfrac{-5 + 5\sqrt{17}}{2}$.

Example 3

Figure 5.3

Find *BD* in Figure 5.3.

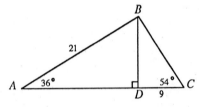

Solution

It is clear that $m\angle ABC = 90°$, since the sum of the measures of the angles of a triangle is $180°$, and so $m\angle ABC = 180° - 36° - 54°$. Thus, we may apply Theorem 5.4. The following is the step-by-step solution (the reader should be able to supply the reasons for each step):

1. $\dfrac{AC}{AB} = \dfrac{AB}{AD}$

2. $\dfrac{AD + 9}{21} = \dfrac{21}{AD}$

3. $(AD)^2 + 9(AD) = 441$

4. $(AD)^2 + 9(AD) - 441 = 0$

5. $AD = \dfrac{-9 + 3\sqrt{205}}{2}$

6. $\dfrac{AD}{BD} = \dfrac{BD}{CD}$

7. $(BD)^2 = AD \cdot CD$

8. $BD = \sqrt{AD \cdot CD}$

9. $BD = 3 \sqrt{\dfrac{-9 + 3\sqrt{205}}{2}}$

10. $BD = \dfrac{3}{2}\sqrt{-18 + 6\sqrt{205}}$

EXERCISE SET 5.3

Figure for Exercises 1–7

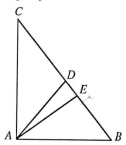

Figure for Exercise 8

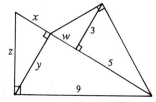

Figure for Exercises 9–10

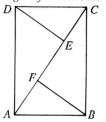

Figure for Exercise 11

In Exercises 1–7, given $\triangle ABC$, $\overline{CA} \perp \overline{AB}$, $\overline{AE} \perp \overline{BC}$, $CD = BD$.

In Exercises 1–3, if $AE = 9$ and $BE = 6$, find:

1. CE

2. AD

3. DE

In Exercises 4–6, if $CE = 5$ and $CA = 8$, find:

4. BE

5. BA

6. AE

7. If $\triangle ABD$ is equilateral and $AB = 4$, find AE.

8. Solve for w, x, y, and z in the figure.

In Exercises 9–10, given: rectangle $ABCD$, $\overline{DE} \perp \overline{AC}$, $\overline{BF} \perp \overline{AC}$.

9. List all similar triangles. Give reasons.

10. List all congruent triangles. Give reasons.

11. Given: $\overline{AB} \parallel \overline{CD}$, $\overline{BD} \perp \overline{CD}$, E is the midpoint of $\overline{AD}$, $BD = AE$

    Prove: $\triangle BED$ is equilateral

5.4 THE PYTHAGOREAN THEOREM AND APPLICATIONS

The following theorem is one of the most famous theorems in plane geometry. There are many proofs of the theorem, as indicated in the historical note in the beginning of this chapter. The proof we use here follows easily from Theorem 5.4. Several more examples of proofs, some of which require the concept of area, are presented in Exercise Set 7.2, No. 32–35.

Theorem 5.5	(Pythagorean Theorem) The square of the length of the hypotenuse of a right triangle equals the sum of the squares of the lengths of the legs.

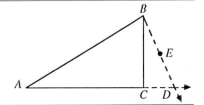

Given: $\triangle ABC$, $\angle C$ is a right $\angle$

Prove: $(AB)^2 = (AC)^2 + (BC)^2$

Proof:

Statements	Reasons
1. $\triangle ABC$, $\angle C$ is a right $\angle$	1. Given
2. There is a ray $\overrightarrow{BE}$ such that E is on the same side of $\overleftrightarrow{AB}$ as C and m$\angle ABE = 90°$. Let $\overrightarrow{BE} \cap \overrightarrow{AC} - \{D\}$.	2. Protractor Post.
3. $\dfrac{AC}{AB} = \dfrac{AB}{AD}$	3. Each leg of a right $\triangle$ is the geom. mean of the hyp. and the seg. of the hyp. adj. to the leg
4. $AD = AC + CD$	4. Def. of between
5. $\dfrac{AC}{AB} = \dfrac{AB}{AC + CD}$	5. Substitution
6. $(AB)^2 = AC(AC + CD)$	6. Mult. Axiom of =
7. $(AB)^2 = (AC)^2 + AC \cdot CD$	7. Distributivity
8. $\dfrac{AC}{BC} = \dfrac{BC}{CD}$	8. The alt. of a right $\triangle$ is the geom. mean of the segments of the hyp.
9. $CD = \dfrac{(BC)^2}{AC}$	9. Mult. Axiom of =
10. $(AB)^2 = (AC)^2 + (BC)^2$	10. Substitution (State. 7 and 9)

∎

Figure 5.4

Example 1

Find AC in Figure 5.4 if $AB = 5$ and $BC = 7$.

Solution

By Theorem 5.5, $(AC)^2 = (AB)^2 + (BC)^2 = 25 + 49 = 74$. Therefore, $AC = \sqrt{74}$.

∎

Example 2

Find AB in Figure 5.4 if $AC = 19$ and $BC = 18$.

Solution

By Theorem 5.5, $(AC)^2 = (AB)^2 + (BC)^2$, from which it follows that $(AB)^2 = (AC)^2 - (BC)^2 = 361 - 324 = 37$. Therefore, $AB = \sqrt{37}$.

∎

The proof of the following corollary to Theorem 5.5 is left as an exercise.

Corollary | In an isosceles right triangle, the hypotenuse is $\sqrt{2}$ times as long as each of the legs.

Example 3

Find the length of each leg of an isosceles right triangle if the hypotenuse has length 12.

Solution

By the corollary to Theorem 5.5, $\sqrt{2}$ times the length of each leg will equal 12. Thus, the length of each leg equals $12/\sqrt{2} = 6\sqrt{2}$.

∎

The following theorem and corollary illustrate some of the applications of the Pythagorean Theorem.

Theorem 5.6 | In a right triangle, if an acute angle has a measure of $30°$, then the leg opposite this angle is half as long as the hypotenuse.

Given: $m\angle C = 90°$, $m\angle A = 30°$

Prove: $BC = AB/2$

Proof:

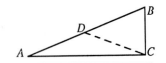

Statements	Reasons
1. $m\angle C = 90°$, $m\angle A = 30°$	1. Given
2. $m\angle B = 60°$	2. The sum of the meas. of the $\angle$s of a $\triangle$ is 180° and Add. Axiom of $=$
3. There is a point D on $\overline{AB}$ such that $AD = DB$	3. Ruler Post.
4. $CD = AB/2$	4. The median to the hyp. of any right $\triangle$ is ½ as long as the hyp.
5. $AB = AD + DB$	5. Def. of between
6. $AB = 2DB$	6. Substitution
7. $CD = DB$	7. Substitution (State. 4 and 6)
8. $m\angle DCB = 60°$	8. If 2 sides of a $\triangle$ are $\simeq$, the $\angle$s opp. these sides are $\simeq$
9. $m\angle BDC = 60°$	9. Same as Reason 2
10. $BC = DB$	10. If 2 $\angle$s of a $\triangle$ are $\simeq$, the sides opp. the $\angle$s are $\simeq$
11. $AB = 2BC$	11. Substitution
12. $BC = AB/2$	12. Mult. Axioms of equality and symmetry

Corollary

> In a right triangle, if an acute angle has a measure of 60°, then the leg opposite this angle is $\sqrt{3}/2$ times as long as the hypotenuse and $\sqrt{3}$ times as long as the other leg.

Given: $m\angle C = 90°$, $m\angle B = 60°$

Prove: $AC = \sqrt{3}AB/2$, $AC = \sqrt{3}BC$

Proof: See Exercise Set 5.4, No. 31.

Figure 5.5

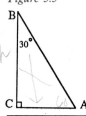

Example 4

In Figure 5.5, find *AC* and *BC* if *AB* = 12.

Solution

By Theorem 5.6, *AC* = 12/2 = 6.
By the corollary to Theorem 5.6, *BC* = 6$\sqrt{3}$.

■

Example 5

In Figure 5.6, find *FG*.

Figure 5.6

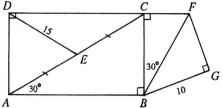

Solution

Since *DE* = 15 and $\overline{DE}$ is a median, we find by Theorem 5.2 that
AC = 30. Hence, by Theorem 5.6, *BC* = 15. Thus, by the corollary to
Theorem 5.6, since *BC* = $\sqrt{3}BF/2$, we find that *BF* = 30/$\sqrt{3}$ = 10$\sqrt{3}$. But
by the Pythagorean Theorem, it follows that $(BF)^2 = (FG)^2 + (BG)^2$, and
so $(FG)^2 = (BF)^2 - (BG)^2 = (10\sqrt{3})^2 - 10^2 = 300 - 100 = 200$. It follows
that *FG* = $\sqrt{200}$ = 10$\sqrt{2}$.

■

We should emphasize that in a computational problem, listing the
reasons is not necessary; but, of course, we must understand how each
step is derived. It is for this reason that we refer to the theorems in the
solution to Example 5.

The following theorem is the converse of the Pythagorean Theorem.
Its proof is left as an exercise.

Theorem
5.7

If the sum of the squares of the lengths of two sides of a triangle
equals the square of the length of the third side, then the triangle is
a right triangle, with the right angle opposite the third side.

EXERCISE SET 5.4

In Exercises 1–4, given: $\triangle PQR$, $\overline{QS} \perp \overline{PR}$, $\overline{PQ} \perp \overline{RQ}$. Find:

Figure for Exercises 1–4

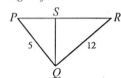

1. *PR*

2. *PS*

3. *RS*

4. *QS*

In Exercises 5–10, find the length of the hypotenuse of each of the right triangles whose legs are given.

5. 1, 2

6. 3, 4

7. 6, 8 ·

8. 5, 12

9. 5, 10

10. $\sqrt{5}/2$, 3

In Exercises 11–16, find the length of the second leg of each of the right triangles, given one leg and the hypotenuse.

11. $L = 4$, $H = 5$

12. $L = 2$, $H = 7$

13. $L = 5$, $H = 5\sqrt{2}$

14. $L = 2/3$, $H = 7/3$

15. $L = 1.2$, $H = 3.4$

16. $L = 17$, $H = 16$

Figure for Exercises 17–22

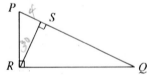

In Exercises 17–22, given: $\triangle ABC$, $\triangle ABD$, $B-C-D$, $m \angle BAC = 30°$, $m \angle ADB = 45°$. Find:

17. *BC*

18. *CD*

19. *AC*

20. *AD*

21. $m \angle DAC$

22. $m \angle ACD$

Figure for Exercises 23–26

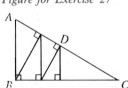

In Exercises 23–26 given: $\triangle PQR$, $\overline{PR} \perp \overline{RQ}$, $\overline{RS} \perp \overline{PQ}$, $m \angle PRS = 30°$, $PS = 4$. Find:

23. $m \angle Q$

24. $m \angle P$

25. *RS*

26. *QS*

Figure for Exercise 27

27. Find *AB* in the figure if $CD = 10$ and $m \angle C = 30°$.

28. Prove Theorem 5.7. (Hint: Construct a right triangle whose legs are congruent to the two shortest sides of the given triangle. Show that the right triangle constructed is congruent to the original triangle.)

29. Prove the corollary to Theorem 5.5.

30. Prove that in a right triangle, if the leg opposite an acute angle is half as long as the hypotenuse, then the angle has a measure of 30°.

31. Prove the corollary to Theorem 5.6.

32. A rectangular vacant lot is 100 feet by 50 feet. How much distance will a dog save by walking along a diagonal of the lot to get to the opposite vertex, rather than walking along the outside of the lot?

33. A rectangular parking lot is 500 feet by 300 feet. How much distance will a car save by driving along a diagonal of the lot to get to the opposite vertex, rather than driving along the outside of the lot?

34. A 30-foot vertical antenna is to be supported by a wire tied to a hook on the ground 20 feet from the base of the antenna. If we assume that 1 extra foot of wire is needed at each end to make a knot, how long a piece of wire is required to join the top of the antenna to the hook?

35. A 70-foot vertical pole is to be supported by a wire tied to a hook on the ground 30 feet from the base of the pole. If we assume that 1 extra foot of wire is needed at each end to make a knot, how long a piece of wire is required to join the top of the pole to the hook?

5.5 TRIGONOMETRY (Optional)

In the last section, we discuss several specific right triangles in which we are able to determine the ratios between the sides. For example, in Theorem 5.6 we prove that in a right triangle with $m\angle A = 30°$, the leg opposite $\angle A$ is half as long as the hypotenuse. Thus, the ratio of the leg opposite $\angle A$ to the hypotenuse is 1/2. In the corollary of Theorem 5.6, we show that if $m\angle B = 60°$, then the leg opposite $\angle B$ is $\sqrt{3}/2$ times as long as the hypotenuse. Thus, the ratio of the leg opposite $\angle B$ to the hypotenuse is $\sqrt{3}/2$. In this section, we generalize the concept of such ratios to other right triangles.

In general, **trigonometry** is the study of the measurement of triangles. This section is intended to be only a short introduction to the subject, since a typical course in trigonometry requires a comprehensive background in algebra.

For the sake of brevity, when we refer to the length of the hypotenuse of a right triangle, we use the abbreviation *HYP*. When we refer to the length of the side opposite an acute angle of a right triangle, we use the abbreviation *OPP*. When we refer to the length of the leg of a right triangle adjacent to an acute angle, we use the abbreviation *ADJ*. Figure 5.7 illustrates the names of the sides of a right triangle relative to ∠A and relative to ∠B.

Figure 5.7

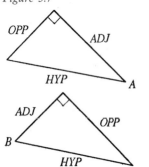

Definition

In a right triangle with acute angle ∠A:

sine m∠A = *OPP/HYP*
cosine m∠A = *ADJ/HYP*
tangent m∠A = *OPP/ADJ*

The three ratios above are called *trigonometric functions* of m∠A and are denoted by sin m∠A, cos m∠A, and tan m∠A, but are pronounced as if the words were not abbreviated. There are three other commonly used trigonometric functions that are the reciprocals of the above three functions. We do not introduce those functions here.

Example 1

Evaluate sin 30°, cos 30°, and tan 30°.

Solution

By Theorem 5.6, sin 30° = 1/2. By the corollary to Theorem 5.6, cos 30° = $\sqrt{3}/2$ and tan 30° = $1/\sqrt{3}$.

In Example 1, we used the fact that if an acute angle of a right triangle has a measure of 30°, then the other acute angle has a measure of 60°. Thus, the leg opposite the angle with measure 60° is adjacent to the angle with measure 30°, and, for example, sin 60° = cos 30°.

Example 2

Evaluate sin 45°, cos 45°, and tan 45°.

Solution

By the corollary to Theorem 5.5, sin 45° = $1/\sqrt{2}$ and cos 45° = $1/\sqrt{2}$. Since the triangle is isosceles, tan 45° = 1.

∎

Example 3

Consider $\triangle ABC$ with $AB = 3$, $AC = 4$, and $BC = 5$. Evaluate sin m∠C, cos m∠C, tan m∠C. See Figure 5.8.

Figure 5.8

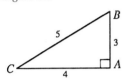

Solution

By Theorem 5.7, $\triangle ABC$ is a right triangle with m∠A = 90°, since $3^2 + 4^2 = 5^2$. Thus, sin m∠C = 3/5, cos m∠C = 4/5, and tan m∠C = 3/4.

∎

It is possible to compute the trigonometric functions of the measures of all angles, but the discussion in this text is limited to acute angles. The theorems that have been discussed facilitate the computation of these functions for specific examples. More advanced textbooks describe methods for finding the trigonometric functions for all possible measures of angles.

The table in Appendix H lists the approximate values of the functions of all acute angles whose measures are rounded to the nearest degree. Precise tables are available, as well as tables in which the angles are measured in radians instead of degrees. Calculators that yield the values of trigonometric functions are also available.

Example 4

Evaluate sin 27°, cos 27°, and tan 27°.

Solution

From Appendix H, we find that sin 27° = .4540, cos 27° = .8910, and tan 27° = .5095.

∎

Example 5

Prove that tan $m\angle A$ = (sin $m\angle A$)/(cos $m\angle A$).

Solution

$$\frac{\sin m\angle A}{\cos m\angle A} = \frac{OPP/HYP}{ADJ/HYP}$$

$$= \frac{OPP}{HYP} \cdot \frac{HYP}{ADJ}$$

$$= \frac{OPP}{ADJ}$$

$$= \tan m\angle A$$

We use trigonometric functions in many applications. Some common applications are demonstrated here. Appendix H is used to evaluate the trigonometric functions.

Example 6

A building casts a shadow 100 feet long when the sun is 65° above the horizon. How tall is the building?

Solution

Figure 5.9

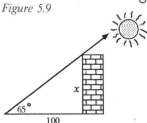

From the diagram in Figure 5.9, it follows that

$$\tan 65° = x/100.$$

Multiplying both sides of this equation by 100 yields

$$x = \tan 65° = (100)(2.1445) = 214.45.$$

so the building is approximately 214 feet tall.

Example 7

During a flight of the space shuttle, an astronaut examines a spherical asteroid known to have a diameter of 2 miles (Figure 5-10). The astronaut notices that the diameter of the asteroid subtends an angle of measure 2°. How far away from the astronaut is the center of the asteroid?

Figure 5.10

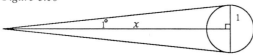

Solution

Since the diameter of the asteroid is two miles, its radius is one mile. Since the asteroid subtends an angle of measure 2°, a right triangle with an acute angle of measure 1° is formed as shown in Figure 5.10. Thus,

$$\tan 1° = 1/x.$$

Multiplying both sides of this equation by x yields

$$x \tan 1° = 1.$$

Dividing both sides of the equation by $\tan 1°$, we get

$$x = 1/(\tan 1°) = 1/(.0175) = 57 \text{ miles}.$$

■

■

Example 8

Alan and Barbara, who are standing 400 meters apart, are throwing rocks at a target across a deep ravine. In Figure 5.11, given $\triangle ABC$ with $m\angle A = 35°$, and $m\angle B = 55°$, Alan is standing at A, Barbara is standing at B, and the target is at C. Find the distances from Alan and Barbara to the target.

Figure 5.11

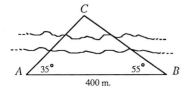

Solution

Notice that $\triangle ABC$ is a right angle since the measures of $\angle A$ and $\angle B$ add up to 90°, and 180° − 90° = 90°. Thus,

$$\cos 35° = AC/400$$

and

$$\sin 35° = BC/400.$$

It follows that

$$AC = 400 \cos 35° = (400)(.8192) = 328 \text{ meters}$$

and

$$BC = 400 \sin 35° = (400)(.5736) = 229 \text{ meters}.$$

■

EXERCISE SET 5.5

In Exercises 1–9, evaluate the trigonometric functions without using tables.

1. $\sin 60°$

2. $\cos 60°$

3. $\sin 45°$

4. $\cos 45°$

5. $\sin 30°$

6. $\cos 30°$

7. $\tan 60°$

8. $\tan 45°$

9. $\tan 30°$

In Exercises 10–15, if $\triangle ABC$ is given with $AB = 8$, $BC = 15$, and $AC = 17$, find:

10. $\sin m\angle A$

11. $\cos m\angle A$

12. $\tan m\angle A$

13. $\sin m\angle C$

14. $\cos m\angle C$

15. $\tan m\angle C$

In Exercises 16–21, if $\triangle ABC$ is given with $m\angle C = 90°$, $AC = 2$, and $BC = 3$, find:

16. $\sin m\angle A$

17. $\cos m\angle A$

18. $\tan m\angle A$

19. $\sin m\angle B$

20. $\cos m\angle B$

21. $\tan m\angle B$

In Exercises 22–33, evaluate the trigonometric functions by using the table in Appendix H.

22. $\sin 17°$

23. $\cos 17°$

24. $\tan 17°$

25. $\sin 79°$

26. $\cos 79°$

27. $\tan 79°$

28. $\sin 33°$

29. $\cos 33°$

30. $\tan 33°$

31. $\sin 86°$

32. $\cos 86°$

33. $\tan 86°$

In Exercises 34–37, evaluate the trigonometric functions by using the table in Appendix H.

34. $\sin 13°$ and $\cos 77°$

35. $\sin 49°$ and $\cos 41°$

36. $\sin 71°$ and $\cos 19°$

37. $\sin 89°$ and $\cos 1°$

38. A building casts a shadow 500 feet long when the sun is 37° above the horizon. How tall is the building?

39. A vertical telephone pole casts a shadow 10 feet long when the sun is 80° above the horizon. How tall is the telephone pole?

40. A basketball player who is 7 feet tall casts a shadow when the sun is 5° above the horizon. How long is the shadow?

41. A 3000-foot vertical cliff casts a shadow upon level ground when the sun is 11° above the horizon. How long is the shadow?

42. A rod supports a vertical televison antenna that is 17 feet tall. The rod makes an angle of measure 34° with the horizontal. How long is the rod?

43. A ladder just reaches a window ledge 10 feet above the horizontal ground. The ladder makes an angle of measure 70° with the ground. How long is the ladder?

44. A spherical balloon known to have a diameter of 10 meters is observed by a woman on a desert island. The woman notices that the diameter of the balloon subtends an angle of measure 6°. How far away from the woman is the center of the balloon?

45. The diameter of a comet 100,000 miles away from a giant is observed by the giant to subtend an angle of measure 2°. What is the radius of the comet?

46. A piece of tile on a kitchen floor is in the shape of a right triangle. A spider is sitting at the vertex C of the right angle and a fly and a gnat occupy the other two vertices, A and B, respectively. If $m\angle A = 22°$, find the distances from the spider to the fly and from the spider to the gnat, if the fly is 8 inches from the gnat.

47. A telephone company worker on top of a vertical 50-foot telephone pole looks down toward a bear on the horizontal ground at an angle of 15° with the horizontal. How far away from the worker is the bear? How far from the base of the pole is the bear?

Prove the statements in Exercises 48–50.

48. $\sin m\angle A = (\cos m\angle A)(\tan m\angle A)$

49. $\cos m\angle B = (\sin m\angle B)/(\tan m\angle B)$

50. $(\sin m\angle A)^2 + (\cos m\angle A)^2 = 1$

Chapter 5
Key Terms

Chapter 5
Review Exercises

Fill each blank with the word *always, sometimes,* or *never*.

1. In the quadratic equation $px^2 + rx + t = 0$, r is _____ zero.

2. A right triangle _____ has two acute angles.

3. In a right triangle, the altitude to the hypotenuse is _____ half as long as the hypotenuse.

4. The hypotenuse of an isosceles right triangle is _____ one of the congruent sides.

5. The altitude to the hypotenuse of a right triangle is _____ shorter than either of the legs.

6. The median to the hypotenuse of a right triangle is _____ shorter than the altitude to the hypotenuse.

7. If a triangle has sides of lengths $2a$, $3a$, and $4a$, it is _____ a right triangle.

8. An isosceles right triangle is _____ similar to any other isosceles right triangle.

9. If $\triangle ABC$ and $\triangle ABD$ are right triangles with right angle at B, and if $m\angle BAC = 30°$ and $m\angle BAD = 45°$, then $\overline{BC}$ is _____ longer than $\overline{BD}$.

10. An altitude of a right triangle _____ forms two similar triangles.

True-False: If the statement is true, mark it so. If it is false, replace the underlined word to make a true statement.

11. If one acute angle of a right triangle is congruent to an acute angle of a second right triangle, the triangles are <u>congruent</u>.

12. The HA = HA congruence theorem for right triangles is a special case of the <u>ASA = ASA</u> congruence theorem.

13. A line segment drawn from a vertex of a rectangle <u>perpendicular</u> to a diagonal is the geometric mean of the two <u>segments of the diagonal</u>.

14. In an isosceles right triangle, the hypotenuse is $\sqrt{3}$ times as long as each of the legs.

15. The median to the hypotenuse of a right triangle forms two isosceles triangles.

16. In a right triangle, if an acute angle has a measure of $60°$, then the leg opposite this angle is twice as long as the leg opposite the angle whose measure is $30°$.

17. If a leg and an acute angle of one right triangle are congruent to a leg and an acute angle of another right triangle, respectively, then the two triangles are not necessarily congruent.

18. The square of the length of the hypotenuse of a right triangle equals the square of the sum of the lengths of the legs.

19. In an isosceles right triangle, the altitude to the hypotenuse forms two right triangles that are congruent to each other.

20. In an isosceles right triangle, the altitude to the hypotenuse forms two isosceles right triangles.

In Exercises 21-23, given: $\triangle ABC$, $\overline{CB} \perp \overline{AB}$, $\overline{BM}$ is a median.

Figure for Exercises 21–23

21. If $BM = 3$, find AC.

22. If $m\angle BAC = 30°$ and $AB = 2$, find BM.

23. If $AB = 2BC$, find AM.

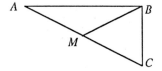

In Exercises 24-27, given: $\square ABDE$, $\triangle ACD$, $\overline{AD} \perp \overline{CD}$, $A–B–C$, $CD = 3$. Find:

Figure for Exercises 24–27

24. BC

25. BD

26. The perimeter of $\triangle ABD$

27. The perimeter of $\square ABDE$

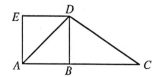

In Exercises 28-31, given: $\triangle ABC$, $\overline{AC} \perp \overline{AB}$, $\overline{AD} \perp \overline{BC}$, $\overline{EF} \parallel \overline{AC}$, $A–D–F$, $D–E–B$, $EF/AC = 3/2$. Find:

Figure for Exercises 28–31

28. DE

29. AD

30. EF

31. EB

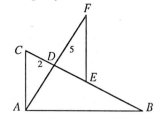

In Exercises 32–37, find the length of the hypotenuse of the right triangles whose legs are given.

32. 10, 24

33. 5, $\sqrt{11}$

34. $\sqrt{2}$, $\sqrt{3}$

35. 8, 15

36. $2\sqrt{2}$, $3\sqrt{3}$

37. 4a, 5a

In Exercises 38–43, find the length of the second leg of each of the right triangles, given one leg and the hypotenuse.

38. $L = 6$, $H = 10$

39. $L = 5$, $H = 8$

40. $L = 2.4$, $H = 2.5$

41. $L = 5/12$, $H = 7/12$

42. $L = \sqrt{3}$, $H = \sqrt{8}$

43. $L = 4n$, $H = 7n$

A *Pythagorean triple* is any set of integers satisfying the equation $x^2 + y^2 = z^2$. For example { 3, 4, 5 } is a Pythagorean triple. In Exercises 44–45, find other Pythagorean triples: (Hint: Let a and b be positive integers so that $a > b$. Let $x = 2ab$, $y = a^2 - b^2$, and $z = a^2 + b^2$. It is easy to show that for these choices { x, y, z } is a Pythagorean triple.)

44. That are multiples of 3, 4, 5.

45. That are not multiples of 3, 4, 5.

In Exercises 46–48, given: $\triangle RST$, $\overline{RS} \perp \overline{ST}$.

46. If $m\angle T = 45°$ and $RT = 12$, find ST.

47. If $m\angle T = 2m\angle R$ and $RS = \sqrt{3}/4$, find RT.

48. If $\overline{SM} \perp \overline{RT}$, $R–M–T$, $RM/MT = 2/3$, and $ST = 4$, find MT.

In Exercises 49–51, given: $\triangle DEF$, $\overline{EF} \perp \overline{DF}$, $\overline{FG} \perp \overline{DE}$, $D–G–E$.

49. If $m\angle D = 37°$, find $m\angle EFG$.

50. If $DG/DF = 3/5$ and $GF = 15$, find EF.

51. If GE, GF, EF is proportional to 3, 4, 5, find GE/DE.

52. Given that rectangle $MNPQ$, $m\angle MPN = 30°$, and $MP = 10$, find the perimeter of the rectangle.

53. Find the perimeter of a regular hexagon whose apothem has a length of $\sqrt{15}$.

54. Find the perimeter of an equilateral triangle whose apothem has length 60 light years.

Figure for Exercises 49–51

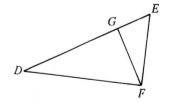

55. Find the apothem of an equilateral triangle with a side of length 5 parsecs.

56. Find the apothem of a regular hexagon whose perimeter is 10 meters.

57. Find the apothem of a regular hexagon with a side of length $\sqrt{3}$ centimeters.

Proofs:

58. *Given: $A-B-C$, $A-E-D$ in the diagram*
Prove: $\triangle CDE \sim \triangle AEB$

Figure for Exercise 58
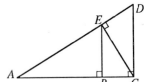

In Exercises 59–60, given: $\triangle MNP$, $\overline{NQ} \perp \overline{MP}$, $MN = NP$, $\overline{QR}$ is an altitude of $\triangle MNQ$, $\overline{QS}$ is an altitude of $\triangle PNQ$. Prove:

Figure for Exercises 59–60

59. $\triangle QRN \simeq \triangle QSN$

60. $SP/SQ = RQ/RN = QP/QN$

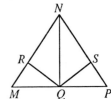

61. Prove that if the apothem of a regular hexagon is x, then the perimeter of the hexagon is $4\sqrt{3}x$.

62. Prove that if the perimeter of an equilateral triangle is y, then the apothem of the triangle is $\sqrt{3}y/18$.

63. *Given:* rectangle *DEGF*,
    $A-D-E$, $D-E-B$, $A-F-C$, $B-G-C$,
    $AE = BD$

Prove: $\triangle ABC$ is isosceles

Figure for Exercise 63
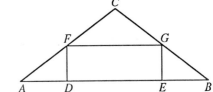

Constructions:

64. Construct a right triangle, given the hypotenuse and an acute angle.

65. Construct a right triangle, given the hypotenuse and a leg.

66. Construct a right triangle, given one leg and the altitude to the hypotenuse. (Hint: Construct two parallel lines so that the distance between them equals the length of the given altitude.)

67. Construct a square, given its diagonal.

In Exercises 68–69, given a line segment one unit long (say 2 inches), construct:

68. A line segment of length $\sqrt{2}$ units

69. A line segment of length $\sqrt{3}$ units

Circles

Felix Klein (1849-1925)

Courtesy of Historical Pictures Service, Chicago

Historical Note

By using a rather sophisticated definition of *distance*, the British mathematician Arthur Cayley (1821–1895) and the German mathematician Felix Klein (1849–1925) formed a theory that unified geometrical concepts. Klein once gave a series of lectures at Göttingen for the purpose of bringing current mathematical ideas to the secondary schools. His lectures covered the three classical problems of elementary geometry.

1. *The duplication of a cube:*
   Given the length of the edge of a cube, construct the length of the edge of a cube of double the volume.

2. *The trisection of an angle:*
   Divide an angle into three congruent angles by construction.

3. *The quadrature of a circle:*
   Construct a square having the same area as a given circle.

In the past, many mathematicians tried to solve them, but the construction of these three problems has been proved impossible. These proofs require algebraic methods.

In this chapter, we discuss certain relationships between circles, lines, and line segments and also introduce the concept of *loci*. Although Section 3.2 introduced the paragraph form, its use there was primarily restricted to indirect proof. Starting with this chapter, we change style by writing proofs in paragraph form. The reader should now be ready to write all proofs in this form.

6.1   TANGENTS

In Section 1.2, we define *circle*, *center*, and *radius*. We use the word *radius* to refer to the line segment, rather than the length of the line segment.

Definition

> A **tangent line** is a line in the plane of a circle, which intersects the circle in exactly one point. This point is called the **point of tangency**.

Definition

> A **tangent ray** is a ray contained in a tangent line, having its endpoint on the point of tangency.

Definition

> A **tangent segment** is a segment contained in a tangent line, having one of its endpoints on the point of tangency.

We often shorten the phrases *tangent line*, *tangent ray*, and *tangent segment* to *tangent*.

■

Example 1

In Figure 6.1, given circle Q ($\odot Q$), (a) name the tangent line; (b) name the tangent rays; and (c) name the tangent segments.

Solution

Figure 6.1

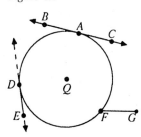

(a) $\overleftrightarrow{BC}$ is a tangent line.

(b) $\overrightarrow{AB}$ and $\overrightarrow{AC}$ are tangent rays.

(c) $\overline{AB}$, $\overline{AC}$, and $\overline{DE}$ are tangent segments.

Note that $\overline{FG}$ is not a tangent segment, since it is not contained in a tangent line.

■

Definition

> The **interior of a circle** P (int $\odot P$) with radius r is the set of all points Q such that $PQ < r$. The **exterior of a circle** P (ext $\odot P$) with radius r is the set of all points Q such that $PQ > r$.

Figure 6.2

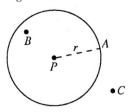

In Figure 6.2, point A is in $\odot P$, point B is in int $\odot P$, and point C is in ext $\odot P$. Clearly, the center of a circle is in the interior of the circle.

Circles in the same plane, having the same center, are **concentric circles**. Two circles in the same plane are **internally tangent** if they intersect in exactly one point and the intersection of their interiors is not empty. Two circles in the same plane are **externally tangent** if they intersect in exactly one point and the intersection of their interiors is empty.

■

Example 2

In Figure 6.3, name the internally tangent circles and the externally tangent circles.

Figure 6.3

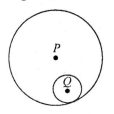
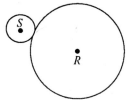

Solution

Circles P and Q are internally tangent and circles R and S are externally tangent.

◼

A tangent of two circles is a **common internal tangent** if the intersection of the tangent and the line segment joining the centers is not empty. A tangent of two circles is a **common external tangent** if the intersection of the tangent and the line segment joining the centers is empty.

◼

Example 3

In Figure 6.4, name the common external tangents and the common internal tangents.

Solution

$\overleftrightarrow{AB}$ and $\overleftrightarrow{CD}$ are common external tangents, and $\overleftrightarrow{EF}$ and $\overleftrightarrow{GH}$ are common internal tangents.

Figure 6.4

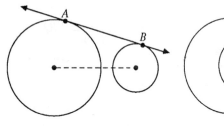 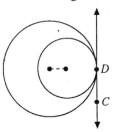 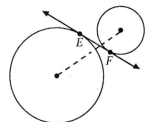 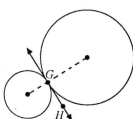

◼

We now consider some of the properties of circles and their tangents. The following theorem shows us how a tangent of a circle is related to the radius of a circle.

Theorem 6.1	A line perpendicular to a radius at a point on a circle is a tangent of the circle.

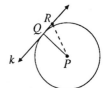

Given: Point $Q \in \odot P$,
          $Q \in$ line k,
          $\overline{PQ} \perp k$

Prove: Line k is a tangent of $\odot P$

Proof: Suppose k is not a tangent of $\odot P$. Then, since $Q \in (k \cap \odot P)$, k must also contain another point R on $\odot P$. By Theorem 2.7, since $\overline{PQ} \perp k$, $\overline{PR}$ is not perpendicular to k. By Theorem 3.11, since $m\angle PQR = 90°$, $m\angle PRQ$ is not greater than $90°$. Hence, $m\angle PRQ < 90°$. It follows, by Theorem 3.3, that $PQ < PR$. But this contradicts the statement that R is on $\odot P$, since $\overline{PQ}$ is a radius of $\odot P$. Therefore, k is a tangent of $\odot P$.

The following theorem is the converse of Theorem 6.1.

Theorem 6.2 | A tangent of a circle is perpendicular to the radius of the circle with an endpoint on the point of tangency.

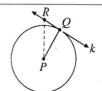

Given: k is a tangent of $\odot P$ at Q

Prove: $\overline{PQ} \perp k$

Proof: Suppose $\overline{PQ}$ is not perpendicular to k. Then, by Theorem 2.7, there is a point R on k such that $\overline{PR} \perp k$. Since $m\angle PRQ = 90°$, $m\angle PQR < 90°$. Thus, by Theorem 3.3, $PR < PQ$. This means that R is in int $\odot P$. Clearly, if R is in int $\odot P$ and line k contains R, k must intersect $\odot P$ in two points. (See Exercise Set 6.1, No. 19.) This contradiction establishes that $\overline{PQ} \perp k$.

A circle is **inscribed in a polygon** if all the sides of the polygon lie on tangents of the circle. We also say that the **polygon circumscribes the circle.**

In Figure 6.5, $\odot P$ is inscribed in quadrilateral *ABCD* and quadrilateral *ABCD* circumscribes $\odot P$.

Figure 6.5

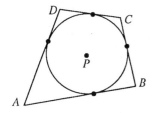

■

Example 4

A space shuttle is in orbit 100 miles above the earth, which has a radius of approximately 4000 miles. Assuming that a cross section of the earth is circular, find the distance that an astronaut in the shuttle can see to the horizon.

Solution

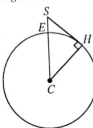

Figure 6.6

In Figure 6.6, $CH = CE = 4000$ miles and $ES = 100$ miles. But $CS = CE + ES = 4100$ miles. By the Pythagorean Theorem,

$$(CS)^2 = (CH)^2 + (HS)^2.$$

Thus,
$$\begin{aligned}(HS)^2 &= (CS)^2 - (CH)^2 \\ &= (4100)^2 - (4000)^2 \\ &= 810000.\end{aligned}$$

Therefore,
$$\begin{aligned}HS &= \sqrt{810000} \\ &= 900 \text{ miles.}\end{aligned}$$

■

EXERCISE SET 6.1

Figure for Exercises 1–11

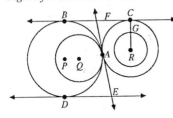

In Exercises 1–11, P, Q, and R are centers of the circles and A, B, C, and D are points of tangency.

In Exercises 1–6, name:

1. $\overline{AF}$ 4. $\overline{RC}$

2. $\overleftrightarrow{BC}$ 5. $\overline{AQ}$

3. $\overrightarrow{DE}$ 6. $\overleftrightarrow{AE}$

In Exercises 7–11, what is the relationship of each of the pairs of circles?

7. $\odot P$ and $\odot Q$

8. $\odot P$ and $\odot R$, such that $C \in \odot R$

9. $\odot R$, such that $C \in \odot R$ and $\odot R$, such that $G \in \odot R$

10. $\odot Q$ and $\odot R$, such that $C \in \odot R$

11. $\odot Q$ and $\odot R$, such that $G \in \odot R$

12. Find the radius of the circle inscribed in a square whose diagonal is 6 units long.

13. If the quadrilateral $ABCD$ circumscribes $\odot P$, $\overline{BC} \perp \overline{AB}$, $\overline{CD} \perp \overline{AD}$, $m\angle C = 120°$, and $PG = 2$, find the perimeter of $ABCD$.

14. *Given:* $\odot P$ and $\odot Q$ intersect at points A and B
    Prove: $\triangle APQ \cong \triangle BPQ$

Figure for Exercise 13

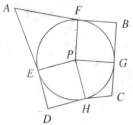

Figure for Exercise 14

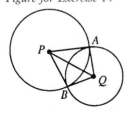

In Exercises 15–18, given $\overleftrightarrow{AC}$ and $\overleftrightarrow{BC}$ are tangents of $\odot P$ and $\odot Q$ and $P-Q-C$. Prove:

Figure for Exercises 15–18

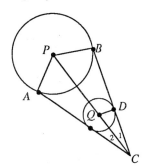

15. $\overline{BC} \simeq \overline{AC}$

16. $\angle 1 \simeq \angle 2$

17. $\overline{PB} \parallel \overline{QD}$

18. $\triangle CDQ \sim \triangle CBP$

19. Prove that a line containing a point in the interior of a circle must intersect the circle in two points.

20. Draw a fairly large scalene triangle and inscribe a circle in it.

21. A space shuttle is in orbit 200 miles above the earth, which has a radius of approximately 4000 miles. Assuming that a cross section of the earth is circular, find the distance an astronaut in the shuttle can see to the horizon.

22. Answer the question in Exercise 21 if the space shuttle is 500 miles above the earth.

23. A spacecraft is in orbit 100 miles above a planet, which has a radius of approximately 2200 miles. Assuming that a cross section of the planet is circular, find the distance an astronaut in the spacecraft can see to the horizon.

24. Answer the question in Exercise 23 if the spacecraft is 300 miles above the planet.

In Exercises 25–28, the sun shining upon the earth causes the earth to cast a shadow. The *umbra* is the part of the shadow included between the common external tangents. The *penumbra* is included between the common internal tangents.

25. Draw a diagram showing the sun and the earth as described, shading the umbra.

26. Draw a diagram showing the sun and the earth as described, shading the penumbra.

27. In which region of the shadow would a total eclipse of the moon occur?

28. In which region of the shadow would a partial eclipse of the moon occur?

6.2 CHORDS AND SECANTS

We now extend the concept of circle relationships by introducing lines and line segments that are not tangents of the circle.

Figure 6.7

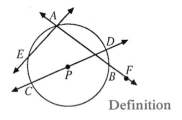

| Definition | A **chord** of a circle is a line segment with endpoints on the circle. A **diameter** is a chord that contains the center of the circle. |

We also use the word *diameter* to refer to the length of the chord involved. The diameter of a circle is twice the radius. In Figure 6.7, given $\odot P$, we see that $\overline{AB}$, $\overline{AE}$, and $\overline{CD}$ are chords, and $\overline{CD}$ is a diameter.

| Definition | A **secant** of a circle is a line that intersects the circle in exactly two points. A **secant ray** is a ray that intersects a circle in exactly two points. |

In Figure 6.7, $\overleftrightarrow{AE}$, $\overleftrightarrow{AB}$, and $\overleftrightarrow{CD}$ are secants, and $\overrightarrow{AE}$, $\overrightarrow{CD}$, and $\overrightarrow{FA}$ are secant rays. (Can you name other secant rays?)

| Theorem 6.3 | If a secant containing the center of a circle is perpendicular to a chord, then it bisects the chord. |

Given: $\odot P$, chord $\overline{AB}$, $\overline{PQ} \perp \overline{AB}$

Prove: $AQ = BQ$

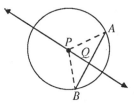

Proof: Consider $\triangle APQ$ and $\triangle BPQ$. Clearly, $AP = BP$ since $\overline{AP}$ and $\overline{BP}$ are radii of $\odot P$. Since $PQ = PQ$ and $\overline{PQ} \perp \overline{AB}$, we have $\triangle APQ \simeq \triangle BPQ$ by HL = HL. Since corresponding parts of congruent triangles are congruent, $AQ = BQ$.

The following converse of Theorem 6.3 is also true. Its proof is left as an exercise.

| Theorem 6.4 | If a secant containing the center of a circle bisects a chord that is not a diameter, then the secant is perpendicular to the chord. |

Theorem 6.5 leads us to the construction of a circle containing three given points not all on one line. Its proof is left as an exercise.

| Theorem 6.5 | In the plane of a circle, the perpendicular bisector of a chord contains the center of the circle. |

| Construction 6.1 | Construct a circle containing three given points not all on one line. |

Given: Points *A*, *B*, and *C*

Construct: ⊙*P* such that {*A, B, C*} ⊂ ⊙*P*

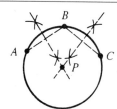

Step 1. Draw $\overline{AB}$ and $\overline{BC}$.

Step 2. Using Construction 1.3, construct the perpendicular bisector of $\overline{AB}$ and of $\overline{BC}$. Since *A*, *B*, and *C* are not all on one line, the bisectors are not parallel, and, thus, intersect at a point *P*.

Step 3. Using *P* as a center and *AP* as a radius, construct ⊙*P*. Then, ⊙*P* will contain *A*, *B*, and *C*, and so ⊙*P* is the required circle.

Two circles are **congruent** if they have equal radii. The following theorems give the relationships between the lengths of chords and their distances from the center of a circle.

| Theorem 6.6 | In the same circle or in congruent circles, chords equidistant from the center are congruent. |

Given: ⊙*P* ≃ ⊙*S*, *PQ* = *ST*, $\overline{PQ} \perp \overline{AB}$, $\overline{ST} \perp \overline{CD}$,

Prove: *AB* = *CD*

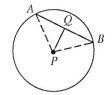 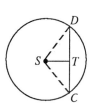

Note: We prove the statement for congruent circles. The proof for the same circle is similar.

Proof: Since ⊙*P* ≃ ⊙*S*, we know that *AP* = *BP* = *CS* = *DS*. As we are given $\overline{PQ} \perp \overline{AB}$, $\overline{ST} \perp \overline{CD}$, and *PQ* = *ST*, it follows from HL = HL that △*APQ* ≃ △*BPQ* ≃ △*CST* ≃ △*DST*. Because corresponding parts of congruent triangles are congruent, $AQ = BQ = CT = DT$. It easily follows that $AB = CD$.

Theorem 6.7

In the same circle or in congruent circles, any two congruent chords are equidistant from the center.

Given: $\odot P \simeq \odot S$,

$AB = CD$,

$\overline{PQ} \perp \overline{AB}$,

$\overline{ST} \perp \overline{CD}$

Prove: $PQ = ST$

Proof: Since $\odot P \simeq \odot S$, we know that $AP = CS$. By Theorem 6.3, we know that $\overline{PQ}$ bisects $\overline{AB}$ and $\overline{ST}$ bisects $\overline{CD}$. Since $AB = CD$, it follows that $AQ = CT$. Since $\overline{PQ} \perp \overline{AB}$ and $\overline{ST} \perp \overline{CD}$, we have $\triangle APQ \simeq \triangle CST$ by HL = HL. Because corresponding parts of congruent triangles are congruent, $PQ = ST$.

Theorem 6.8

Given two chords in the same circle or in congruent circles, the chord nearer the center is longer.

Given: $\odot P \simeq \odot S$,

$\overline{PQ} \perp \overline{AB}$,

$\overline{ST} \perp \overline{CD}$,

$PQ < ST$

Prove: $AB > CD$

Proof: By the Pythagorean Theorem, $(AQ)^2 + (PQ)^2 = (AP)^2$ and $(CT)^2 + (ST)^2 = (CS)^2$. Since $\odot P \simeq \odot S$, $AP = CS$. Thus, $(AQ)^2 + (PQ)^2 = (CT)^2 + (ST)^2$. We are given that $PQ < ST$; hence, $(AQ)^2 + (ST)^2 > (AQ)^2 + (PQ)^2$. This implies that $(AQ)^2 + (ST)^2 > (CT)^2 + (ST)^2$. Thus, $(AQ)^2 > (CT)^2$, so that $AQ > CT$ and hence, $2AQ > 2CT$. But by Theorem 6.3, $AQ = BQ$ and $CT = DT$, which implies that $AB = 2AQ$ and $CD = 2CT$. Therefore, $AB > CD$.

A **polygon is inscribed in a circle** if all the sides of the polygon are chords of the circle. We also say that the **circle circumscribes a polygon.** In Figure 6.8, pentagon $ABCDE$ is inscribed in $\odot P$ and $\odot P$ circumscribes pentagon $ABCDE$.

Figure 6.8

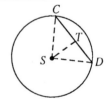

EXERCISE SET 6.2

Figure for Exercises 1–4

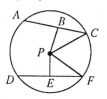

In Exercises 1–4, given $\odot P \simeq \odot R$, $EF = AB$. **Insert the correct symbol (=, <, or >).**

1. RT _____ PQ

2. CD _____ EF

3. GS _____ AP

4. GH _____ CD

5. Find the perimeter of the inscribed quadrilateral $EFGH$ in Exercise Set 6.1, No. 13.

6. Draw a circle and mark three points A, B, and C on it, not on a straight line. Find the center of the circle by construction.

7. Draw a triangle and circumscribe a circle about it.

8. Inscribe a regular hexagon in a given circle.

9. *Given:* $\odot P \simeq \odot Q$, secant k containing P and Q, secant m containing the point A of intersection of the circles, $\overline{BC} \parallel \overline{PQ}$
   Prove: $m\angle ABP = m\angle ACQ$

Figure for Exercise 9

Figure for Exercise 10

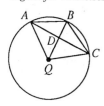

10. *Given:* $\odot P$, chords $\overline{AC}$ and $\overline{DF}$, $\overline{BP} \perp \overline{AC}$, $\overline{EP} \perp \overline{DF}$, and $BP = PE$
    Prove: $\angle C \simeq \angle F$

In Exercises 11–12, given $\{A, B, C\} \subset \odot Q$, $\overline{BQ} \perp \overline{AC}$. **Prove:**

11. $AB = BC$

12. $m\angle CAB = m\angle ACB$

Figure for Exercises 11–12

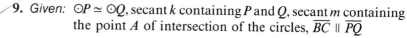

13. Prove that exactly one circle can be drawn through any three points not on a line. (Hint: Prove that such a circle exists and no other circle contains the same three points.)

14. Prove that no circle contains three different points on a line.

15. Prove Theorem 6.4.

16. Prove Theorem 6.5.

6.3 ARC-ANGLE RELATIONSHIPS

In this section, we discuss relationships between the arcs and angles of a circle. We begin with descriptions of a semicircle and a central angle.

Definition

> A **semicircle** of a circle P is the union of the endpoints of its diameter with the set of all points of the circle on one side of the diameter.

Example 1

Name the semicircles shown in Figure 6.9.

Figure 6.9

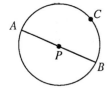

Solution

One semicircle is $\{A, B\} \cup \{Q: Q$ is on $\odot P$ and on the same side of $\overleftrightarrow{AB}$ as $C\}$.

Another semicircle is $\{A, B\} \cup \{Q: Q$ is on $\odot P$ and on the opposite side of $\overleftrightarrow{AB}$ from $C\}$.

■

Definition

> A **central angle** of a circle is an angle whose vertex is the center of the circle.

Figure 6.10

In Figure 6.10, $\angle P$ is a central angle of $\odot P$.

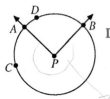

Definition

> A **minor arc** AB of a circle P with points A and B on $\odot P$ and $m\angle APB \neq 180°$ is the union of A and B with the set of all points of $\odot P$ in int$\angle APB$.

Definition

> A **major arc** ACB of a circle P with points A and B on $\odot P$, $m\angle APB \neq 180°$, and $C \in (\odot P \cap $ ext$\angle APB)$ is the union of A and B with the set of all points of $\odot P$ in ext$\angle APB$.

Definition | An **arc** is either a semicircle, a minor arc, or a major arc. We write $\widehat{AB}$ or $\widehat{ACB}$ to represent arc AB and arc ACB, respectively.

In Figure 6.10, the points A and B are *endpoints* of the arcs $\widehat{AB}$ and $\widehat{ACB}$. We say that $\widehat{AB}$ or $\widehat{ACB}$ are **intercepted** by $\angle APB$. Notationally, we always use three letters to denote a major arc, but we may use two or three letters for a minor arc, depending on the need for clarity. Thus, in Figure 6.10, $\widehat{AB} = \widehat{ADB}$.

Definition | The **measure of a minor arc** is the measure of the central angle containing the endpoints of the arc. We write $\mathrm{m}\widehat{AB}$ for the measure of minor arc $\widehat{AB}$.

Definition | The **measure of a major arc** is 360° minus the measure of the central angle containing the endpoints of the arc. We write $\mathrm{m}\widehat{ACB}$ for the measure of major arc $\widehat{ACB}$.

■

Example 2

In Figure 6.10, find $\mathrm{m}\widehat{AB}$ and $\mathrm{m}\widehat{ACB}$ if $\mathrm{m}\angle APB = 87°$.

Solution

We have $\mathrm{m}\widehat{AB} = 87°$ and $\mathrm{m}\widehat{ACB} = 360° - 87° = 273°$.

■

The proof of the following theorem is left as an exercise.

Theorem 6.9 | **(Arc Addition Theorem)** If C is a point of $\widehat{AB}$, then $\mathrm{m}\widehat{AB} = \mathrm{m}\widehat{AC} + \mathrm{m}\widehat{CB}$.

Given: $\{A, B, C\} \subset \odot P$

Prove: $\mathrm{m}\widehat{AB} = \mathrm{m}\widehat{AC} + \mathrm{m}\widehat{CB}$

Proof: See Exercise Set 6.3, No. 24.

Example 3

Figure 6.11

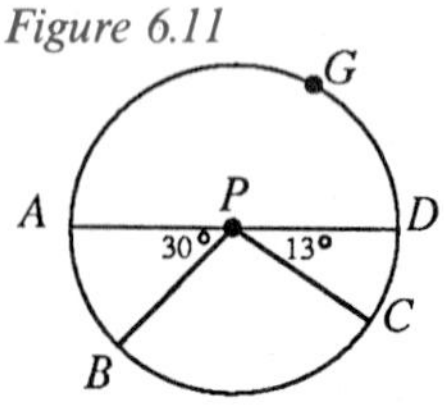

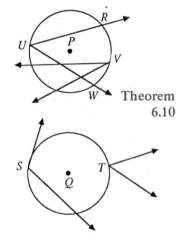

If $\overline{AD}$ is a diameter in Figure 6.11, find $m\widehat{AGD}$, $m\widehat{BC}$, $m\widehat{BD}$, and $m\widehat{BAD}$.

Solution

$m\widehat{AGD} = 180°$, $m\widehat{BC} = 137°$, $m\widehat{BD} = 150°$, and $m\widehat{BAD} = 210°$.

Definition

An **inscribed angle** of a circle P is an angle whose vertex is on circle P and whose rays each intersect the circle in a point other than the vertex.

Figure 6.12

In Figure 6.12, $\angle U$ and $\angle V$ are inscribed angles, but $\angle S$ and $\angle T$ are not inscribed angles. Why? In Figure 6.12, we say that $\angle RUW$ is *inscribed in* $\widehat{RUW}$.

Theorem 6.10

The measure of an inscribed angle is half the measure of its intercepted arc.

Given: $\{A, B, C\} \subset \odot P$

Prove: $m\angle BAC = m\widehat{BC}/2$

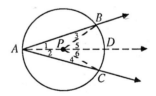

Note: There are three cases to consider. Case 1: $P \in \text{int}\angle BAC$, Case 2: $P \in \angle BAC$, and Case 3: $P \in \text{ext}\angle BAC$. We consider only Case 1 here. Cases 2 and 3 are left as exercises.

Proof: We draw $\overrightarrow{AP}$ intersecting $\odot P$ at a new point D. We draw $\overline{PB}$ and $\overline{PC}$. By Theorem 3.12, $m\angle 5 = m\angle 1 + m\angle 3$ and $m\angle 6 = m\angle 2 + m\angle 4$. But $AP = BP = CP$, so that by Theorem 2.4, $m\angle 1 = m\angle 3$ and $m\angle 2 = m\angle 4$. Thus, $m\angle 5 = 2m\angle 1$ and $m\angle 6 = 2m\angle 2$. By the definition of the measure of arc, $m\widehat{BD} = m\angle 5$ and $m\widehat{CD} = m\angle 6$. Thus, $m\widehat{BD} + m\widehat{DC} = 2m\angle 1 + 2m\angle 2 = 2(m\angle 1 + m\angle 2)$. By the Arc Addition Theorem and the Angle Addition Postulate, $m\widehat{BC} = 2m\angle BAC$. Thus, $m\angle BAC = m\widehat{BC}/2$.

The proofs of the following corollaries are left as exercises.

Corollary 1 | Any inscribed angle intercepting a semicircle is a right angle.

Corollary 2 | All inscribed angles of a circle intercepting the same arc are congruent.

The following theorem is closely related to Theorem 6.10.

Theorem 6.11 | The measure of an angle whose vertex is on a circle, such that one ray is a tangent and the other a secant, equals half the measure of its intercepted arc.

Given: $\{A, C, D\} \subset \odot P, A-P-D$,
$D \in \text{int} \angle BAC, \overrightarrow{AB}$ is a
tangent of $\odot P$

Prove: $m\angle BAC = m\widehat{ADC}/2$

Note: There are two other cases to consider, namely $D \in \text{ext} \angle BAC$ and $D \in \angle BAC$.

Proof: See Exercise Set 6.3, No. 29.

The following two theorems are proved by using Theorem 6.10.

Theorem 6.12 | Given a pair of vertical angles with the vertex in the interior of a circle, the measure of each angle equals half the sum of the measures of the two intercepted arcs.

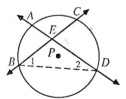

Given: $\{A, B, C, D\} \subset \odot P$

Prove: $m\angle AEB = m\angle CED$
$= (m\widehat{AB} + m\widehat{CD})/2$

Proof: Since vertical angles are congruent, $m\angle AEB = m\angle CED$. By Theorem 6.10, $m\angle 1 = m\widehat{CD}/2$ and $m\angle 2 = m\widehat{AB}/2$. By Theorem 3.12, $m\angle AEB = m\angle 2 + m\angle 1$. Therefore, $m\angle AEB = (m\widehat{AB} + m\widehat{CD})/2$.

Theorem
6.13

Given an angle with the vertex in the exterior of a circle, such that each of the rays of the angle intersect the circle, the measure of the angle equals half the absolute value of the difference of the measures of the intercepted arcs.

Given: $\{A, B, C, D\} \subset \odot P$

Prove: $m\angle C = (m\widehat{AE} - m\widehat{BD})/2$

Proof: See Exercise Set 6.3, No. 30.

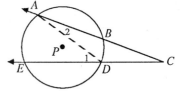

By taking the absolute value, we ensure that the result is positive, and we need not concern ourselves with the order in which the arcs are subtracted. Since $m\widehat{AE} > m\widehat{BD}$ from the diagram, we see that $|m\widehat{AE} - m\widehat{BD}| = m\widehat{AE} - m\widehat{BD}$.

The diagrams in Figure 6.13 represent other cases of Theorem 6.13. (See Exercise Set 6.3, No. 31.) For simplicity, we state Theorem 6.13 as justification for any statements involving the situations occurring in Figure 6.13.

Figure 6.13

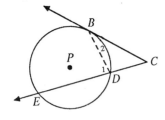
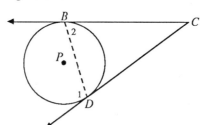

Example 4

In Figure 6.14, find $m\widehat{BC}$, $m\angle ACB$, and $m\angle BDC$.

Figure 6.14

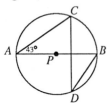

Solution

Using Theorem 6.10, we find that $m\widehat{BC} = 86°$.
By Corollary 1, $m\angle ACB = 90°$.
By Corollary 2, $m\angle BDC = 43°$.

Example 5

Figure 6.15

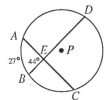

In Figure 6.15, find $m\widehat{CD}$.

Solution

By Theorem 6.12, $m\angle AEB = (m\widehat{AB} + m\widehat{CD})/2$.
Thus, $m\widehat{CD} + 27° = 2(44°)$, so that $m\widehat{CD} = 88° - 27° = 61°$.

Example 6

In Figure 6.16, find $m\angle C$.

Figure 6.16

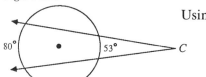

Solution

Using Theorem 6.13, we find that
$$m\angle C = (80° - 53°)/2 = 13.5°.$$

EXERCISE SET 6.3

1. List all the minor arcs in the figure for Exercise Set 6.1, No. 13.

2. List all the major arcs in the figure for Exercise Set 6.1, No. 13.

3. Give reasons for the statements referring to Figure 6.12.

Figure for Exercises 4–5

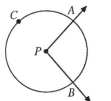

In Exercises 4–5, given $m\angle BPA = 94°$ in $\odot P$, find:

4. $m\widehat{AB}$

5. $m\widehat{ACB}$

In Exercises 6–9, given $\overline{AB}$ is a diameter, $m\angle BPC = 27°$, find:

Figure for Exercises 6–9

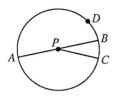

6. $m\widehat{AC}$

7. $m\widehat{ADB}$

8. $m\widehat{ABC}$

9. $m\widehat{BAC}$

In Exercises 10–13, given $m\angle BAC = 30°$, $m\widehat{CD} = 50°$, $\odot P$, find:

Figure for Exercises 10–13

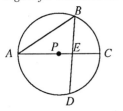

10. $m\widehat{BC}$

11. $m\widehat{AB}$

12. $m\angle BEC$

13. $m\angle ABD$

In Exercises 14–18, given $\overline{AC}$ is a diameter of $\odot P$, t is a tangent of $\odot P$ at E, find:

Figure for Exercises 14–18

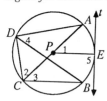

14. $m\widehat{AE}$, if $m\angle 1 = 43°$

15. $m\angle 2$, if $m\angle A = 21°$

16. $m\angle 3$, if $m\angle 4 = 49°$

17. $m\angle 5$

18. $m\widehat{BC}$, if $m\widehat{AD} = 140°$, $m\angle 4 = 45°$, and $m\angle B = 20°$

In Exercises 19–23, given $\overrightarrow{PE}$ is a tangent of $\odot Q$ at E, $\overrightarrow{PA}$ is a tangent of $\odot Q$ at A, find:

Figure for Exercises 19–23

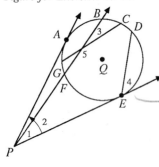

19. $m\angle 3$, if $m\widehat{FG} = 20°$ and $m\widehat{BC} = 30°$

20. $m\widehat{DE}$, if $m\angle 4 = 52°$

21. $m\angle 2$, if $m\widehat{AFE} = 110°$

22. $m\widehat{FE}$, if $m\widehat{BDE} = 160°$ and $m\angle 1 = 35°$

23. $m\widehat{BC}$, if $m\widehat{GF} = 25°$ and $m\angle 5 = 145°$

24. Prove Theorem 6.9. (Hint: There are several cases, such as both minor arcs or one minor arc and one major arc.)

25. Prove Case 2 of Theorem 6.10.

26. Prove Case 3 of Theorem 6.10.

27. Prove Corollary 1 of Theorem 6.10.

28. Prove Corollary 2 of Theorem 6.10.

29. Prove all cases of Theorem 6.11.

30. Prove Theorem 6.13.

31. State and prove theorems comparable to Theorem 6.13 for the diagrams in Figure 6.13.

32. Prove that parallel lines intercept congruent arcs on a circle.

6.4 CIRCLE-SEGMENT RELATIONSHIPS

In this section, we discuss the way in which the lengths of certain segments are determined by their relationships to circles. The proof of Theorem 6.14 is left as an exercise.

Theorem 6.14	In the same circle or in congruent circles, two chords are congruent if and only if the corresponding arcs are congruent.

Theorem 6.15	The two tangent segments from a point to a circle are congruent.

Given: $\{B, C\} \subset \odot P$, $\overline{AB}$ and $\overline{AC}$ are tangents of $\odot P$

Prove: $\overline{AB} \simeq \overline{AC}$

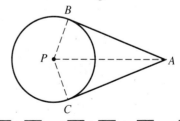

Proof: By Theorem 6.2, $\overline{AB} \perp \overline{BP}$ and $\overline{AC} \perp \overline{CP}$. As $\overline{BP}$ and $\overline{CP}$ are radii of $\odot P$, $BP = CP$. By reflexivity, $AP = AP$. Hence, by HL = HL, $\triangle ABP \simeq \triangle ACP$. Therefore, $\overline{AB} \simeq \overline{AC}$, since corresponding parts of congruent triangles are congruent.

Example 1

In Figure 6.17, if $\overline{AB}$ and $\overline{AC}$ are tangents of $\odot P$, and $AB = 13$, find AC.

Figure 6.17

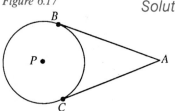

Solution

By Theorem 6.15, $AC = 13$.

Theorem 6.16

If two chords of a circle intersect in the interior of the circle, the product of the lengths of the segments on one chord equals the product of the lengths of the segments on the other chord.

Given: $\{A, B, C, D\} \subset \odot P$

Prove: $AE \cdot DE = BE \cdot CE$

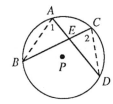

Proof: Since vertical angles are congruent, $m\angle AEB = m\angle CED$. By Corollary 2 of Theorem 6.10, $m\angle 1 = m\angle 2$. Hence, by AA ~ AA (Theorem 4.1), $\triangle ABE \sim \triangle CDE$. Thus, by definition of similar triangles, $AE/CE = BE/DE$. Therefore, $AE \cdot DE = BE \cdot CE$ by the Multiplication Axiom of Equality.

Example 2

Figure 6.18

In Figure 6.18, find BE.

Solution

By Theorem 6.16, we have $BE = 21/5$, since $3 \cdot 7 = 5BE$.

The following two theorems can be proved in much the same way as Theorem 6.16.

Theorem 6.17

If $\{B, C, D, E\} \subset \odot P$ and point A satisfies $A-B-C$ and $A-D-E$, then $AB \cdot AC = AD \cdot AE$.

Proof: See Exercise Set 6.4, No. 15.

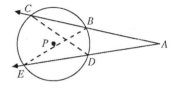

Theorem 6.18

If $\{B, C, D\} \subset \odot P$ and $\overleftrightarrow{AB}$ is a tangent of $\odot P$ where point A satisfies $A-C-D$, then $(AB)^2 = AC \cdot AD$.

Proof: See Exercise Set 6.4, No. 16.

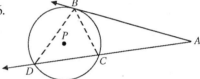

Example 3

Let the circle in the figure for Theorem 6.18 represent a cross-section of the earth. Consider an astronaut who is located in space at point A. He uses a laser device to calculate the distance from his position to cities at points B and C on the earth. He finds that $AB = 9000$ miles and that $AC = 8000$ miles. If a straight telephone cable could be strung from C to D, find its length.

Solution

By Theorem 6.8,

$$(AB)^2 = AC \cdot AD.$$

Thus,

$$
\begin{aligned}
AD &= (AB)^2/AC \\
&= (9000)^2/8000 \\
&= 81000/8 \\
&= 10125 \text{ miles.}
\end{aligned}
$$

Therefore,

$$
\begin{aligned}
CD &= 10125 - 8000 \\
&= 2125 \text{ miles.}
\end{aligned}
$$

EXERCISE SET 6.4

Figure for Exercises 1–2

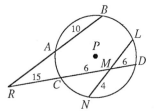

Figure for Exercises 3–6

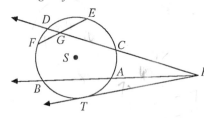

Figure for Exercise 8

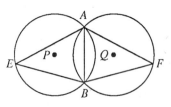

Figure for Exercises 9–10

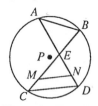

Figure for Exercises 11–12

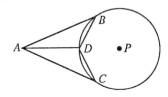

In Exercises 1–2, given ⊙P with lengths as shown, find:

1. *LM*

2. *RA*

In Exercises 3–6, given ⊙S with tangent $\overrightarrow{PT}$ and secants $\overrightarrow{PB}$ and $\overrightarrow{PD}$.

3. If *EG* = 6, *CG* = 12, and *DG* = 2, find *FG* and *EF*.

4. If *PC* = 12, *PA* = 10, and *CD* = 6, find *AB* and *PB*.

5. If *PA* = 12 and *AB* = 4, find *PT*.

6. If *PA* = 6 and *PT* = 7, find *PB*.

7. Construct a line segment, whose length is the geometric mean of the lengths of two given line segments, by applying Theorem 6.18.

8. *Given:* ⊙P ≃ ⊙Q, they intersect at *A* and *B*, and $\widehat{AE} \simeq \widehat{AF}$
   Prove: △*EAB* ≃ △*FAB*

In Exercises 9–10, given *AB* = *MN* and $\overline{MN} \parallel \overline{CD}$, prove:

9. △*ABE* ~ △*CDE*

10. △*ABE* ≃ △*MNE*

In Exercises 11–12, given ⊙P with tangents $\overline{AB}$ and $\overline{AC}$ and *A*−*D*−*P*, prove:

11. ∠*BAD* ≃ ∠*CAD*

12. $\widehat{BD} \simeq \widehat{CD}$

13. Prove that in the same circle or in congruent circles, a chord nearer the center intercepts the greater minor arc.

14. Prove Theorem 6.14.

15. Prove Theorem 6.17.

16. Prove Theorem 6.18.

17. Let the circle in the figure for Theorem 6.18 represent a cross-section of the earth. Consider an astronaut who is located in space at point *A*. He uses a laser device to calculate the distance from his position to cities at points *B* and *C* on the earth. He finds that

AB = 7000 miles and that AC = 5000 miles. If a straight telephone cable could be strung from C to D, find its length.

18. Answer the question in Exercise 17 if the astronaut finds that AB = 8000 miles and that AC = 7000 miles.

6.5 RELATED CONSTRUCTIONS

In this section, we describe how to construct tangents of circles and how to construct a line segment whose length is the geometric mean of the lengths of two given line segments. Using Theorem 6.1, we are able to make the following constructions.

Construction 6.2	Construct a tangent of a circle through a given point of the circle.

Given: Point $A \in \odot P$

Construct: Line k so that k is a tangent of $\odot P$

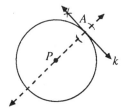

Step 1. Draw $\overleftrightarrow{AP}$.

Step 2. Using Construction 1.5, construct the perpendicular k to $\overleftrightarrow{AP}$ through A. Line k is the required line.

Using Corollary 1 of Theorem 6.10, we are able to make the following constructions.

Construction 6.3	Construct a tangent of a circle through a given point in the exterior of the circle.

Given: Point A is in the exterior of $\odot P$

Construct: $\overleftrightarrow{AB}$ so that $\overleftrightarrow{AB}$ is a tangent of $\odot P$

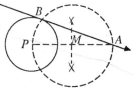

Step 1. Draw $\overline{AP}$, and use Construction 1.3 to find the midpoint M of $\overline{AP}$.

Step 2. Using *M* as the center, construct a circle of radius *MP*. Let one of the points of intersection of this circle with ⊙*P* be labeled *B*.

Step 3. Draw $\overleftrightarrow{AB}$. Then by Corollary 1 of Theorem 6.10 and by Theorem 6.1, $\overleftrightarrow{AB}$ is the required tangent.

Construction 6.4

Construct a common external tangent of two given circles.

Given: ⊙*P* and ⊙*Q* with the radius of ⊙*P* greater than the radius of ⊙*Q*

Construct: $\overleftrightarrow{AB}$ so that $\overleftrightarrow{AB}$ is a tangent of both ⊙*P* and ⊙*Q*, and $\overline{PQ} \cap \overleftrightarrow{AB} = \emptyset$

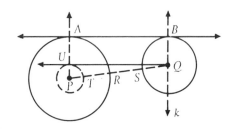

Step 1. Draw $\overline{PQ}$. Let ⊙*P* ∩ $\overline{PQ}$ = {*R*} and ⊙*Q* ∩ $\overline{PQ}$ = {*S*}.

Step 2. Use Construction 1.1 to construct $\overline{RT}$ on $\overline{RP}$ so that *RT* = *SQ*.

Step 3. Construct a circle with center *P* and radius *PT*. Use Construction 6.3 to construct a tangent of this circle from *Q*. Label the point of tangency *U*.

Step 4. Draw $\overrightarrow{PU}$, letting the intersection of $\overrightarrow{PU}$ with the original ⊙*P* be labeled *A*.

Step 5. Use Construction 3.1 to construct a line *k* through *Q* parallel to $\overrightarrow{PU}$. Line *k* intersects ⊙*Q* in two points. One of these points is on the same side of $\overleftrightarrow{PQ}$ as *A*. Label this point *B*, and draw $\overleftrightarrow{AB}$. Then, $\overleftrightarrow{AB}$ is the required line.

The same construction applies when the two given circles intersect. See Exercise Set 6.5, No. 5 for the case when the two circles have equal radii. See Exercise Set 6.5, No. 10 for the proof that the steps in Construction 6.4 provide the desired results.

Construction 6.5

Construct a common internal tangent of two given circles.

Given: ⊙*P* and ⊙*Q* with the radius of ⊙*P* greater than the radius of ⊙*Q*

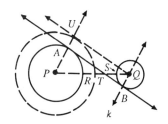

Construct: $\overleftrightarrow{AB}$ so that $\overleftrightarrow{AB}$ is a tangent of
both $\odot P$ and $\odot Q$, and
$\overline{PQ} \cap \overleftrightarrow{AB} \neq \emptyset$

Step 1. Draw $\overline{PQ}$. Let $\odot P \cap \overline{PQ} = \{R\}$ and $\odot Q \cap \overline{PQ} = \{S\}$.

Step 2. Use Construction 1.1 to construct $\overline{RT}$ on $\overline{RS}$ such that
$RT = SQ$.

Step 3. Construct a circle with center P and radius PT. Use Construction 6.3 to construct a tangent of this circle from Q. Label the point of tangency U.

Step 4. Draw $\overrightarrow{PU}$, letting the intersection of $\overrightarrow{PU}$ with the original $\odot P$ be labeled A.

Step 5. Use Construction 3.1 to construct a line k through Q parallel to $\overrightarrow{PU}$. Line k intersects $\odot Q$ in two points. One of these points is not on the same side of $\overleftrightarrow{PQ}$ as A. Label this point B. Draw $\overleftrightarrow{AB}$. Then, $\overleftrightarrow{AB}$ is the required line.

See Exercise Set 6.5, No. 7 for the case when the two circles have equal radii. See Exercise Set 6.5, No. 11 for the proof that the steps in Construction 6.5 provide the desired results.

Construction 6.6	Construct a segment whose length is the geometric mean between the lengths of two given segments.

Given: $\overline{AB}$ and $\overline{CD}$

Construct: $\overline{PR}$ so that $\dfrac{AB}{PR} = \dfrac{PR}{CD}$

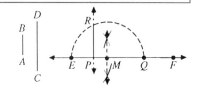

Step 1. Draw $\overleftrightarrow{EF}$. Using Construction 1.1, find a point P on $\overleftrightarrow{EF}$ such that $EP = AB$, and find a point Q so that $E{-}P{-}Q$ and $PQ = CD$.

Step 2. Use Construction 1.3 to find the midpoint M of $\overline{EQ}$. Using M as the center, construct a semicircle with $\overline{EQ}$ as the diameter.

Step 3. Use Construction 1.5 to find the perpendicular to $\overleftrightarrow{EF}$ through P. Let the point of intersection of this perpendicular with the semicircle be R. Then $\overline{PR}$ is the required segment since by Corollary 1 of Theorem 6.10, $\angle ERQ$ is a right angle and since $\overline{EQ} \perp \overline{PR}$ we have $\dfrac{EP}{PR} = \dfrac{PR}{PQ}$

EXERCISE SET 6.5

1. Construct the tangent of a circle through a point on the circle.

2. Construct the two tangents of a circle through a point in the exterior of the circle.

3. Construct a common external tangent of two nonintersecting circles with unequal radii.

4. Construct a common external tangent of two intersecting circles with unequal radii.

5. Construct a common external tangent of two nonintersecting circles with equal radii.

6. Construct a common internal tangent of two nonintersecting circles with unequal radii.

7. Construct a common internal tangent of two nonintersecting circles with equal radii.

8. Construct a line segment whose length is the geometric mean between 3 and 5, using Construction 6.6.

9. Construct a line segment whose length is $\sqrt{3}$.

10. In Construction 6.4, prove that $\overleftrightarrow{AB}$ is a tangent of both $\odot P$ and $\odot Q$.

11. In Construction 6.5, prove that $\overleftrightarrow{AB}$ is a tangent of both $\odot P$ and $\odot Q$.

6.6 LOCI

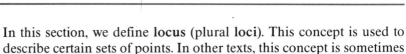

In this section, we define **locus** (plural **loci**). This concept is used to describe certain sets of points. In other texts, this concept is sometimes used to describe the path of a moving point.

Definition | A **locus** is the set of all points that satisfy one or more given conditions.

When given a locus problem, first draw a diagram representing the given conditions. Using the diagram, apply the knowledge gained so far, to ascertain which points satisfy the requirements. Finally, describe the locus in words, naming the type of set formed (point, line, ray, and so on) and where it is located relative to the given set of points. In the following examples, solid lines represent given conditions and dashed lines represent the required loci.

Figure 6.19

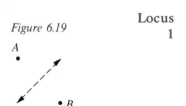

Locus 1	Find the locus of points in a plane equidistant from two given points in the plane.

Given points A and B in Figure 6.19, we find that the locus is the line that is the perpendicular bisector of $\overline{AB}$.

Figure 6.20

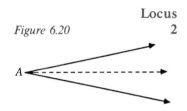

Locus 2	Find the locus of points in the interior of an angle equidistant from the sides of the angle.

Given $\angle A$ in Figure 6.20, we find that the locus is the ray (not including the endpoint) that bisects $\angle A$.

Figure 6.21

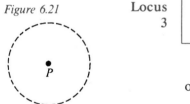

Locus 3	Find the locus of points in a plane that are one unit from a given point in the plane.

Given point P in Figure 6.21, the locus is $\odot P$, which has a radius of one unit.

Figure 6.22

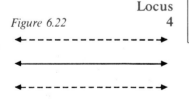

Locus 4	Find the locus of points in a plane that are one unit from a given line in the plane.

Given line k in Figure 6.22, the locus is two parallel lines, each of which is one unit from line k.

Figure 6.23

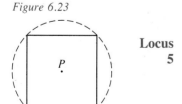

Locus 5	Find the locus of points in a plane that are $\sqrt{2}$ units from the center of a square whose apothem is 1 unit long.

Given the square with center P in Figure 6.23, the locus is the circle which circumscribes the square.

Some locus problems require much analysis, as indicated by the following example.

Locus
6

Find the locus of points in a plane that are one unit from a given point in the plane and half a unit from a given line in the plane.

Figure 6.24

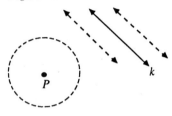

Case 1: The locus is empty. See Figure 6.24.

Figure 6.25

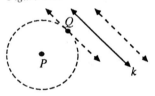

Case 2: The locus is the single point Q. See Figure 6.25.

Figure 6.26

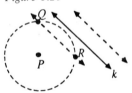

Case 3: The locus is the set of two points $\{Q, R\}$. See Figure 6.26.

Figure 6.27

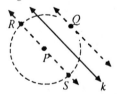

Case 4: The locus is the set of three points $\{Q, R, S, \}$. See Figure 6.27.

Figure 6.28

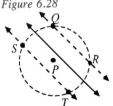

Case 5: The locus is the set of four points $\{Q, R, S, T\}$. See Figure 6.28.

EXERCISE SET 6.6

In this exercise set, assume that all loci are in the plane that contains the given conditions.

1. Construct an acute angle with sides $\overrightarrow{AB}$ and $\overrightarrow{AC}$. Find the locus of points, not in ext$\angle BAC$, equidistant from $\overrightarrow{AB}$ and $\overrightarrow{AC}$.

2. Construct a right angle with sides $\overrightarrow{AB}$ and $\overrightarrow{AC}$. Find the locus of points, not in ext$\angle BAC$, equidistant from $\overrightarrow{AB}$ and $\overrightarrow{AC}$.

3. Let $\overline{AB}$ be the common side of a set of triangles. Find the locus of the vertices opposite $\overline{AB}$ of these triangles, if the altitude of all the triangles is 2 units.

4. Let $\overline{AB}$ be the common side of a set of triangles. Find the locus of the vertices opposite $\overline{AB}$ of these triangles, if the altitude of all the triangles is 1 unit.

5. Find the locus of points equidistant from three given points not on a line.

6. Find the locus of the vertices of all right triangles having $\overline{MN}$ as hypotenuse.

In Exercises 7–8, draw lines m and n 1 inch apart.

7. Find the locus of points equidistant from m and n.

8. Find the locus of points 2 inches from m and 3 inches from n.

9. Find the locus of points 1 inch from the two points A and B. (three cases)

10. Find the locus of points described in Exercise 1 if the points must also be 2 inches from a point P. (How many cases can you find?)

11. Find the locus of the midpoints of a set of parallel chords of a given circle.

12. Find the locus of points that are either 1 inch or 2 inches from a point P.

13. Find the locus of points that are either 1 inch or 2 inches from a circle of radius 3 inches.

14. Find the locus of points at a given distance from a triangle.

15. Find the locus of points at a given distance from a regular pentagon.

Chapter 6
Key Terms

Chapter 6
Review Exercises

Fill each blank with the word *always*, *sometimes*, or *never*.

1. Two circles in a plane _____ have a common tangent.

2. A circle can _~~never~~_ be inscribed in a triangle.

3. A minor arc of a circle _____ has a measure greater than 180°.

4. A diameter of a circle is _____ perpendicular to some chord of the circle.

5. A circle can _____ be constructed that will contain any three points in a plane.

6. A chord of a circle is _____ shorter than a diameter of the circle.

7. The measure of an angle inscribed in a minor arc of a circle is _____ less than 90°.

8. A common internal tangent of two circles in a plane is _____ perpendicular to the line segment joining their centers.

9. If a diameter intersects a circle at A and B and is perpendicular to chord $\overline{CD}$, then the tangents of the circle at A and B are _____ parallel to $\overline{CD}$.

10. Concentric circles _____ have a common tangent.

True-False: If the statement is true, mark it so. If it is false, replace the underlined word to make a true statement.

11. The tangent segment joining the points of tangency of a common external tangent of two circles of unequal radii is <u>shorter</u> than the segment joining the centers of the circles.

12. A point is in the interior of, on, or in the exterior of a circle depending on whether its distance from the center of the circle is <u>greater</u> than, equal to, or <u>less</u> than the radius.

13. A circle <u>circumscribes</u> a polygon if all the sides of the polygon are tangents of the circle.

14. In the same circle or congruent circles, if two major arcs are unequal, the arc of greater measure has the <u>longer</u> chord.

15. If the length of the segment joining the centers of two circles in a plane equals the sum of their radii, the circles must be <u>tangent externally</u>.

16. If chords $\overline{AB}$ and $\overline{CD}$ of circle P intersect at E, then <u>$AE/CE = ED/EB$</u>.

17. The arc of a circle intercepted by an inscribed angle has a measure that is <u>half</u> the measure of the angle.

18. A <u>tangent</u> is a line segment whose endpoints are on a circle.

19. Two chords in the same circle or in congruent circles are congruent if the corresponding <u>major</u> arcs are congruent.

20. A <u>tangent</u> of a circle can contain a chord of the circle.

Answer the following questions.

In Exercises 21–25, $\overrightarrow{AB}$ and $\overrightarrow{AC}$ are tangent rays and $\overrightarrow{AD}$ is a secant ray of $\odot P$.

21. If $AC = 10$, find AB.

22. If $AC = 12$ and $AG = GD$, find AD.

23. If $CF = 3$, $FE = 5$, and $GD = 8$, find FD.

24. If $m\widehat{GBE} = 160°$ and $m\widehat{CD} = 50°$, find $m\angle EFD$.

25. If $m\widehat{CG} = 40°$, $m\widehat{EC} = 65°$, and $m\widehat{ED} = 20°$, find $m\angle CAG$.

Figure for Exercises 21–25

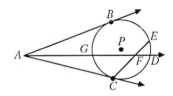

State a reason for each of Exercises 26–33.

In Exercises 26–29, $\overrightarrow{PA}$ and $\overrightarrow{PB}$ are tangent rays of $\odot Q$, $AC = BC$, $\overline{QE} \perp \overline{AC}$ and $\overline{QF} \perp \overline{BC}$.

26. $m\widehat{AC} = m\widehat{BC}$

27. $EQ = QF$

28. $PA = PB$

29. $\overline{QB} \perp \overline{PB}$

Figure for Exercises 26–29

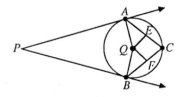

In Exercises 30–33, $\overline{MN}$ is a diameter of $\odot P$, $\overline{KL} \parallel \overline{NM}$, $\overline{PR} \perp \overline{LM}$, and $\overline{MN} \perp \overline{LT}$.

30. $m\widehat{KN} = m\widehat{LM}$

31. $m\widehat{LMN} = m\widehat{MNK}$

32. $LS = ST$

33. $PR > PS$

Figure for Exercises 30–33

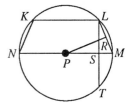

Proof:

34. Make the following construction and prove that the steps provide the desired result: Given three parallel lines, construct an equilateral triangle whose vertices lie on these lines.

Figure for Exercise 34

Construction:

Step 1. Let P be any point on the middle line j.

Step 2. Draw $\overrightarrow{PA}$ and $\overrightarrow{PC}$, each of which makes an angle with j whose measure is $60°$.

Step 3. Through the points A, P, and C, construct a circle Q. Let $\odot Q \cap j = \{B\}$. Then, $\triangle ABC$ is the required triangle.

Loci: In these exercises, assume that all the loci are in the plane that contains the given conditions.

35. Find the locus of points 1 inch from a rectangle whose dimensions are 2 and 3 inches.

36. Find the locus of the midpoints of a set of chords of equal length of a given circle.

37. Find the locus of points that are equidistant from a pair of intersecting lines.

38. Find the locus of points equidistant from a given line and a given point not on the line. (parabola)

39. Find the locus of points of the positions a bug would occupy if the bug were sitting on the spoke of a wheel that is rolling along a flat street (trochoid). Draw a diagram only for the following three cases: (1) the bug on the rim, (2) the bug in the middle of the spoke, and (3) the bug at the center of the wheel.

Constructions:

40. Draw a scalene triangle. Circumscribe a circle about the triangle and inscribe a circle in the triangle.

41. Given A on $\odot P$ and $B \in$ ext $\odot P$, construct a tangent of $\odot P$ at A and a tangent of $\odot P$ from B.

42. Given two nonintersecting circles (in a plane), with unequal radii, construct a common internal tangent and a common external tangent of the circle.

43. Given $\overline{AB}$ and $\overline{CD}$, construct $\overline{DE}$ so that $AB/DE = DE/CD$.

44. Draw a large scalene triangle. Construct the altitudes. Label the points of intersection of the altitudes with the sides A, B, and C. Label the midpoints of the line segments joining the vertices and the point of intersection of the altitudes D, E, and F. Label the midpoints of the sides G, H, and I. Construct the circle containing these points (nine-point circle).

45. Given a circle with an inscribed triangle $\triangle ABC$ and an arbitrary point P on the circle such that P is not a vertex of $\triangle ABC$, construct the three perpendiculars from P to the lines containing the sides of the triangle. The intersection of each of these perpendiculars with one of the lines is called a **foot** of P. If the construction is done accurately, the three feet of P will be contained in a single line, called a **Simson line**.

46. The **centroid** of a triangle is the intersection of its medians. The **orthocenter** of a triangle is the intersection of lines containing the altitudes. The **circumcenter** of a triangle is the center of the circle which circumscribes the triangle. Given $\triangle ABC$, construct its centroid, orthocenter, and circumcenter. If the construction is done accurately, these three points will be contained in a single line, called an **Euler line**.

47. The **incenter** of a triangle is the intersection of its angle bisectors. Construct the incenter of a scalene triangle.

48. If $A - P - B$ and $AB/AP = AP/PB$, then $\overline{AB}$ is divided in the **golden section**. Construct $\overrightarrow{BC} \perp \overline{AB}$. Find Q on $\overrightarrow{BC}$ so that $BQ = AB/2$. Construct $\odot Q$ with radius BQ. Let $\overrightarrow{AQ} \cap \odot Q = \{D, E\}$ where $AD < AE$. If $A - P - B$ such that $AP = AD$, then $AB/AP = AP/PB$.

49. If a line segment is divided in the golden section as described in Exercise 48, and if a rectangle has sides of length equal to AP and PB, then the rectangle is a **golden rectangle**. Construct a golden rectangle from a line segment of length 3 inches.

50. Using the definition from No. 49, construct a golden rectangle $ABCD$ whose longer side AB has a length of 5 inches. Construct another golden rectangle $BCEF$ inside $ABCD$. Construct another golden rectangle $BGHF$ inside $BCEF$. Construct another golden rectangle $FIJH$ inside $BGHF$. Construct another golden rectangle $HKLJ$ inside $FIJH$. Connect the points A, E, G, I, K with a smooth curve as shown in the figure. If you continue in this manner, you will be constructing a **spiral** closely related to ones found in nature. An example is the mollusk known as a nautilus, which has a many-chambered spiral shell.

Figure for Exercise 50

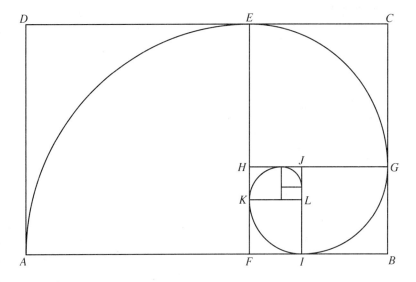

51. The golden section, as described in No. 48, may be used to construct a regular pentagon or a regular decagon inscribed in a circle. Draw a circle with center A and radius $\overline{AB}$, and divide the radius in the golden section at point P. The segment $\overline{BC} \simeq \overline{AP}$ is the side of the regular decagon inscribed in circle A. Connect alternate vertices of the decagon to construct the regular pentagon inscribed in circle A. A five pointed star may be constructed by connecting alternate vertices of the regular pentagon.

52. Construct a regular pentagon circumscribed about a circle. See No. 51.

Areas

Archimedes (287?–212 B.C.)

Courtesy of Historical Pictures Service, Chicago

Historical Note

Archimedes has been called "the Great Geometer." He was particularly interested in practical problems. While bathing, he discovered a method for determining the specific gravity of an object. He is said to have leaped out of the bath and run through the streets naked, yelling "Eureka," (I have found it). He made approximations for π by using geometric and harmonic means. Archimedes estimated π to lie between 3 10/71 and 3 10/70, which is a remarkably good approximation for his time.

So far in this text, we have discussed concepts of measure of line segments and measure of angles, but we have not discussed the measure of regions. In this chapter, we discuss the concept of regions and present postulates and theorems related to areas of regions (regions include those formed by polygons, circles, and portions of circles).

7.1   AREA POSTULATES

In Chapter 1, we defined the interior of a triangle. The interior of a polygon is defined similarly. In Chapter 6, we defined the interior of a circle. Using these definitions of interior, we now define *region*.

Definition | **A polygonal region** is the union of a polygon and its interior.

Definition | **A circular region** is the union of a circle and its interior.

Definition

A region is:
(1) the union of a finite number of polygonal or circular regions, or
(2) the intersection of a finite number of polygonal and circular regions such that the intersection of their interiors is nonempty, or
(3) the union of a finite number of combinations of (1) or (2).

Figure 7.1

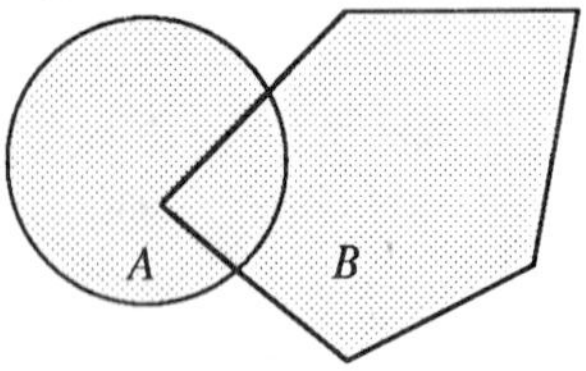

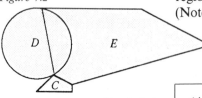

It is possible to give more general definitions of regions, such as infinite regions, but for the purposes of this text, the boxed definition is sufficient. In Figure 7.1, we see that $A \cup B$ is a region, and $A \cap B$ is a region.

In Figure 7.2, although $C \cup (D \cap E)$ is a region, $C \cap (D \cup E)$ is not a region, since $C \cap (D \cup E)$ contains no circular or polygonal region. (Note that $C \cap (D \cup E)$ is a line segment.)

Figure 7.2

Postulate
7.1

(Area Postulate)   To every region there corresponds exactly one positive real number.

The number referred to in the preceding postulate is the **area** of the region. When discussing the area of a particular region, such as a hexagonal region, we say "area of the hexagon," for brevity. We denote the area of polygon *ABCDE* by area*ABCDE* and the area of circle *P* by area⊙*P*.

Postulate 7.2	If two triangles are congruent, then their areas are equal.

■

Example 1

If area*ABC* = 10 and △*ABC* ≃ △*DEF*, find area*DEF*.

Solution

By Postulate 7.2, area△*DEF* = 10.

■

Postulate 7.3	**(Area Addition Postulate)** If the intersection of the interiors of two regions is empty, then the area of the union of the regions equals the sum of the areas of the regions.

■

Example 2

Figure 7.3

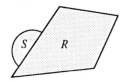

In Figure 7.3 and Figure 7.4, find area($R \cup S$) if area*R* = 17 and area*S* = 12.

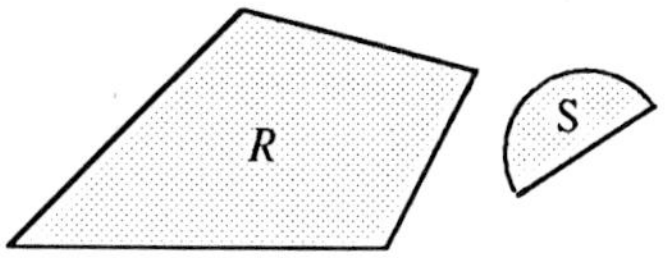

Figure 7.4

Solution

In both cases, by Postulate 7.3,
$$\text{area}(R \cup S) = 17 + 12 = 29.$$

■

Figure 7.5

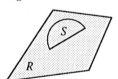

In the statement of the Area Addition Postulate, the intersection of the two interiors of the regions is required to be empty. It should be clear from Figure 7.5 that the area of region R plus the area of region S does not equal the area of region $R \cup S$. In particular, area($R \cup S$) = area$R$. This does not contradict the statement of the Area Addition Postulate, since int$R \cap$ int$S \neq \emptyset$.

■

Example 3

Figure 7.6

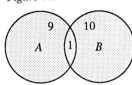

In Figure 7.6, if area$\odot A$ = 10 and area$\odot B$ = 11, and area($A \cap B$) = 1, find area($A \cup B$).

Solution

Area($A \cup B$) = 20. (Don't count the intersecton twice.)

■

Postulate 7.4	**(A = s²)** The area of a square equals the square of the length of a side.

Given a square whose sides are each one unit long, we say the area of the square is 1 **square unit**. For example, a square whose sides are each 1 kilometer long has an area of 1 square kilometer. The area of a square whose sides are each 1 mile long is 1 square mile.

■

Example 4

Find the area of a square that has a side 3 inches long.

Solution

The area of the square is $(3 \text{ in})^2 = 9 \text{ in}^2$.

■

The units in the solution to Example 4 are square inches, written as in², for short.

EXERCISE SET 7.1

Figure for Exercises 1–6

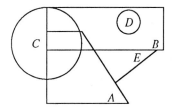

In Exercises 1–6, which of the sets are regions? Why?

1. $C \cup D$

2. $A \cap B$

3. $B \cap E$

4. $(A \cup B) \cap D$

5. $A \cap B \cap C$

6. $(B \cap E) \cup (A \cap E)$

In Exercises 7–12, find the area of each of the regions if area$A = 10$, area$B = 6$, area$C = 12$, and area$(A \cap C) = 2$.

7. $A \cap B$

8. $A \cup B$

9. $B \cup C$

10. $A \cup C$

11. $B \cap A \cap C$

12. $A \cup B \cup C$

Figure for Exercises 7–12

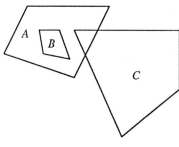

13. *Given:* Square $ABCD$,
    $\overline{EF} \perp \overline{CD}$,
    $DF = CF = 2$ in,
    area$DEF = 7$ in^2
    Find area$ABCED$.

Figure for Exercise 13

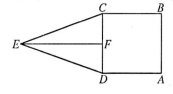

In Exercises 14–17, find the length of a side and of a diagonal of each square whose area is given.

14. 25

15. 18 in^2

16. 7 ft^2

17. 10 cm^2

18. Find the area of a square if the length of its diagonal is $10\sqrt{2}$.

19. If the length of a side of a square is three times the length of a side of a given square, what is the ratio of the area of the given square to the area of the other square?

20. Find the length of a side of a square if the number of feet in its perimeter equals the number of square feet in its area.

21. If a diagonal of one square whose area is 36 is a side of a second square, find the area of the second square.

Figure for Exercise 22

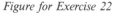

22. Prove that the area of $\square ABCE$ equals the area of $\square ABDF$.

23. Prove that two congruent polygons have equal areas.

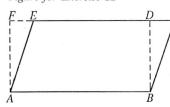

7.2 POLYGONAL REGIONS

In this section, we discuss methods of finding the areas of various polygonal regions. The following theorem is often stated in terms of the length and width of a rectangle, but we prefer the usage shown here in order to prevent ambiguity.

Theorem 7.1

(A = sa)   The area of a rectangle is the product of the length of any side and the length of the altitude to that side.

Given: ▱*ABCD*

Prove: area*ABCD* = (*AB*)(*BC*)

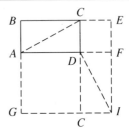

Proof: Let *E, F, G, H,* and *I* be points such that *B−C−E, CE = AB, A−D−F, DF = AB, B−A−G, AG = BC, C−D−H, DH = BC, E−F−I,* and *FI = BC.* Then, clearly, *CEFD, ADHG,* and *BEIG* are squares. Thus, area*CEFD* = (*AB*)^2, area*ADHG* = (*BC*)^2, and

$$\text{area}BEIG = (AB + BC)^2 = (AB)^2 + 2(AB)(BC) + (BC)^2.$$

By Post. 7.2, area*ABC* = area*CDA* = area*IHD* = area*DFI*; thus, by the Area Add. Post., area*ABCD* = area*DFIH.* By the Area Add. Post., we also have

$$\text{area}ABCD + \text{area}CEFD + \text{area}ADHG$$
$$+ \text{area}DFIH = \text{area}BEIG.$$

Hence,

$$2(\text{area}ABCD) + (AB)^2 + (BC)^2 = (AB)^2 + 2(AB)(BC) + (BC)^2.$$

Therefore, area*ABCD* = (*AB*)(*BC*).

■

Example 1

Find the area of a rectangle whose adjacent sides have lengths of 4 inches and 7 inches.

Solution

The area of the rectangle is (4 in)(7 in) = 28 in².

∎

Theorem 7.2	**(A = sa)** The area of a parallelogram is the product of the length of any side and the length of the altitude to that side.

Given: ▱*ABCD*
$\overline{DE} \perp \overline{AB}$

Prove: area*ABCD* = (*AB*)(*DE*)

Proof: Let *F* be a point such that *A*–*B*–*F* and $\overleftrightarrow{CF} \parallel \overleftrightarrow{DE}$. Clearly, $\overline{BF} \perp \overline{CF}$. By Theorem 3.8, $\angle A \simeq \angle CBF$. By the corollary to Theorem 3.15, *AD* = *BC*. Thus, △*ADE* ≃ △*BCF* by HA = HA. Hence, by Post. 7.2, area*ADE* = area*BCF*. By the Area Add. Post., area*ABCD* = area*ADE* + area*DEBC*. Thus, area*ABCD* = area*BCF* + area*DEBC* = area*DEFC*. But *DEFC* is a rectangle and hence, by Theorem 7.1, area*ABCD* = area*DEFC* = (*EF*)(*DE*). Clearly, *EF* = *AB*. Therefore, area*ABCD* = (*AB*)(*DE*).

∎

Example 2

Find the area of the parallelogram in Figure 7.7, if *CD* = 15 feet, *AD* = 7 feet and m∠*DAE* = 60°.

Figure 7.7

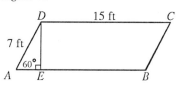

Solution

We find that *AB* = 15 ft, but we need to find *DE*. Using the corollary to Theorem 5.6, we find that *DE* = 7√3/2 ft. Thus,

$$\text{area}ABCD = (15 \text{ ft})(7\sqrt{3}/2 \text{ ft}) = 105\sqrt{3}/2 \text{ ft}^2.$$

∎

Theorem 7.3

(A = sa/2)   The area of a triangle is half the product of the length of any side and the length of the altitude to that side.

Given: $\triangle ABC, \overline{CD} \perp \overline{AB}$

Prove: area $ABC = AB \cdot CD/2$

Proof: See Exercise Set 7.2, No. 26.

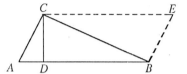

Corollary 1

(A = ap/2)   The area of a regular polygon is half the product of the length of its apothem and its perimeter.

Proof: See Exercise Set 7.2, No. 27.

Corollary 2

If in two triangles one pair of corresponding sides are congruent and the altitudes to these sides are congruent, then the areas of the triangles are equal.

Proof: See Exercise Set 7.2, No. 28.

Definition

An **altitude of a trapezoid** is a segment perpendicular to the two bases of the trapezoid, with endpoints on the base.

Theorem 7.4

[A = a(b₁ + b₂)/2]   The area of a trapezoid is half the product of the length of its altitude and the sum of the lengths of its bases.

Given: $ABCD$ with $\overline{AB} \parallel \overline{CD}$, $\overline{DE} \perp \overline{AB}$

Prove: area $ABCD = DE(AB + CD)/2$

Proof: See Exercise Set 7.2, No. 29.

Note: In the formula of Theorem 7.4, b_1 is the length of one base and b_2 is the length of the other base.

Example 3

Find the area of a trapezoid with bases of length 4 inches and 7 inches and an altitude of length 6 inches.

Solution

The area of the trapezoid is

$$(6 \text{ in})(4 \text{ in} + 7 \text{ in})/2 = (6 \text{ in})(11 \text{ in})/2 = 33 \text{ in}^2.$$

Theorem 7.5

If two triangles are similar, then the ratio of their areas equals the square of the ratio of the lengths of any two corresponding sides *or* the square of the ratio of the lengths of any two corresponding altitudes.

Given: $\triangle ABC \sim \triangle DEF$
$\overline{CG} \perp \overline{AB}, \overline{FH} \perp \overline{DE}$

Prove: $\dfrac{\text{area} ABC}{\text{area} DEF} = \left(\dfrac{AB}{DE}\right)^2 = \left(\dfrac{CG}{FH}\right)^2$

Proof: According to Theorem 7.3, $\text{area} ABC = AB \cdot CG/2$ and $\text{area} DEF = DE \cdot FH/2$. By Exercise 4.3, No. 13, $\dfrac{CG}{FH} = \dfrac{AB}{DE}$.

Therefore,

$$\frac{\text{area} ABC}{\text{area} DEF} = \frac{AB \cdot CG/2}{DE \cdot FH/2} = \frac{AB}{DE} \cdot \frac{CG}{FH} = \frac{AB}{DE} \cdot \frac{AB}{DE} = \left(\frac{AB}{DE}\right)^2.$$

Also,

$$\frac{AB}{DE} \cdot \frac{CG}{FH} = \frac{CG}{FH} \cdot \frac{CG}{FH} = \left(\frac{CG}{FH}\right)^2.$$

Example 4

If $\triangle ABC \sim \triangle DEF$, $\text{area} DEF = 29$ square inches, and $AB/DE = 2/3$, find the area of $\triangle ABC$.

Solution

By Theorem 7.5,

$$\frac{\text{area}\,ABC}{\text{area}\,DEF} = \frac{\text{area}\,ABC}{29\text{ in}^2} = \frac{4}{9}, \text{so that area}\,ABC = \frac{116}{9}\text{ in}^2.$$

∎

Construction 7.1

Construct a triangle equal in area to a given polygon.

Given: Polygon *ABCDE*

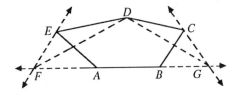

Construct: △*DFG* so that area*DFG* = area*ABCDE*

Step 1. Draw $\overleftrightarrow{AB}$. Construct a parallel to $\overleftrightarrow{AD}$ through *E* and a parallel to $\overleftrightarrow{BD}$ through *C* using Construction 3.1. Let *F* and *G* be the respective intersection points of the parallel lines with $\overleftrightarrow{AB}$.

Step 2. Draw $\overline{DF}$ and $\overline{DG}$. △*DFG* is the required triangle since area*BCD* = area*BDG* and area*ADE* = area*ADF* (See Exercise Set 7.2, No. 31).

EXERCISE SET 7.2

In Exercises 1–6, find the area of each of the polygons.

1. △*ABC*; $\overline{CD} \perp \overline{AB}$, *A–D–B*, *AB* = 5, *CD* = 4.

2. $\square$*ABCD*; $\overline{DE} \perp \overline{AB}$, *A–E–B*, *AB* = 8, *DE* = *AB*/3.

3. $\square$*ABCD*; AB = 3*CD*, the perimeter of *ABCD* is 40.

4. Trapezoid *ABCD*; $\overline{AB} \parallel \overline{CD}$, $\overline{CE} \perp \overline{AB}$, *A–E–B*, *AB* + *CD* + *CE* = 34, *CE*/*CD* = 1/3, *CD*/*AB* = 2/3.

5. Trapezoid *ABCD*; $\overline{BC} \parallel \overline{AD}$, $\overline{CD} \perp \overline{AD}$, *BC* = 10, *AD* = 12, m∠*A* = 45°.

6. $\triangle KLM$; $\triangle KLM \sim \triangle ABC$, $\overline{KL}$ corresponds to $\overline{AB}$, $AB/KL = 3/10$, area $ABC = 10$.

7. Find the area of one side of a rectangular board 70 inches long and 5 inches wide.

8. Find the area of a football field in square feet if it is 160 feet wide and 100 yards long. What is the area in square yards?

9. If the length of one side of a rectangle is 12 and the area is 90, find the length of an adjacent side.

10. If the area of a rectangle is 150 square feet, and the length of one side is three times the length of an adjacent side, find the dimensions of the rectangle.

11. One can of a certain paint is guaranteed to cover 80 square feet. If the walls and ceiling of a 12- by 15-foot room are to be painted, how many cans of this paint should be purchased? The walls are 8 feet 4 inches high, and there are two windows whose dimensions are each 4 by 6 feet.

12. Answer Exercise 11 if the walls and ceiling of a 20- by 30-foot room are to be painted, and one can of paint is guaranteed to cover 65 square feet.

13. The page of a book is 7.5 inches wide by 9.5 inches high. If the top margin is 1 inch, the bottom margin is 0.75 inches, the right side margin is 0.75 inches, and the left side margin is 2 inches, find the area of the printed portion of the page.

14. Answer Exercise 13 using these measurements. If the page is 10 inches wide by 14 inches high, the top margin is 1.5 inches, the bottom margin is 2.5 inches, and the right and left margins are 1.25 inches each.

15. Find the area of a regular hexagon if the length of its apothem is 12 feet.

16. Find the area of a rhombus if the length of a side is 4 inches and the measure of an angle is 60°.

17. Find the area of a right triangle if the lengths of its hypotenuse and a leg are 13 and 7, respectively.

18. Find the length of the altitude of a trapezoid if the lengths of its bases are 22 feet and 30 feet and its area is 273 feet^2.

Hero's Formula may be used to find the area of a triangle if no altitude of the triangle can be found. In $\triangle ABC$, let

$$s = (AB + BC + CA)/2.$$

Then:

$$\text{area}ABC = \sqrt{s(s - AB)(s - BC)(s - CA)}.$$

Apply Hero's Formula to find the area of the triangle in Exercises 19–20, if the lengths of the sides are given.

19. 6, 7, 11

20. 3, 8, 5

21. Draw a nonregular polygon *ABCDE* and construct a triangle equal in area to the polygon.

22. Draw a nonregular polygon *ABCDEF* and construct a triangle equal in area to the polygon.

In Exercises 23–24, draw a parallelogram and construct a triangle equal in area to the parallelogram.

23. Use Construction 7.1.

24. Find another method of construction using one side of the parallelogram as a side of the triangle.

25. Prove that the area of a right triangle is half the product of the lengths of its legs.

26. Prove Theorem 7.3.

27. Prove Corollary 1 of Theorem 7.3.

28. Prove Corollary 2 of Theorem 7.3.

29. Prove Theorem 7.4.

30. State and prove a theorem for polygons comparable to Theorem 7.5.

31. Prove that the area of the triangle equals the area of the polygon in Construction 7.1.

Many proofs of the Pythagorean Theorem (Theorem 5.5) are based on area. Some of these proofs are outlined in Exercises 32–35.

32. *Given:* $\triangle ABC$, $\angle C$ is a right angle
    Prove: $(AB)^2 = (AC)^2 + (BC)^2$
    Plan of proof: Area $\square ADEB$ equals the sum of the areas of four triangles, each congruent to $\triangle ABC$, plus the area of $\square FGHI$.

Figure for Exercise 32

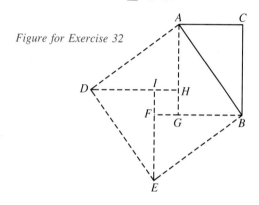

33. (President Garfield's Proof)
    Given: $\triangle ABC$, $\angle C$ is a right angle
    Prove: $(AB)^2 = (AC)^2 + (BC)^2$
    Plan of proof: Let $D - A - C$, $DA = CB$, $\overline{DE} \parallel \overline{CB}$, and $DE = AC$. Then the area of trapezoid $BCDE$ equals the area of $\triangle BAE$ plus twice the area of $\triangle ABC$.

Figure for Exercise 33

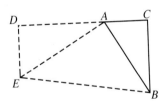

34. *Given:* $\triangle ABC$, $\angle C$ is a right angle
    Prove: $(AB)^2 = (AC)^2 + (BC)^2$
    Plan of proof: Consider circle A and the theorem about segments on intersecting chords.

Figure for Exercise 34

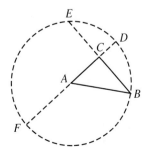

Figure for Exercise 35

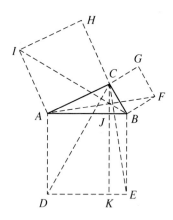

35. (Euclid's Proof)

Given: $\triangle ABC$, $\angle C$ is a right angle

Prove: $(AB)^2 = (AC)^2 + (BC)^2$

Plan of proof: Consider a square on each side of $\triangle ABC$, and $\overline{CK} \perp \overline{DE}$. Area of $\square ABED$ equals the sum of the areas of rectangle $ADKJ$ and rectangle $BEKJ$. The area of rectangle $ADKJ$ equals twice the area of $\triangle ADC$ since they have the same base and altitude. This equals twice the area of $\triangle ABI$ (by congruence), which equals the area of $\square ACHI$. Similar reasoning shows that the area of rectangle $BEKJ$ equals the area of $\square BFGC$.

7.3 CIRCULAR REGIONS, SECTORS, AND SEGMENTS

In this section, we discuss the methods of finding areas of circular regions, sectors, and segments. We begin with definitions of a sector and a segment.

Definition

A **sector** is a region that is the union of:

(a) $\overset{\frown}{ACB}$ of $\odot P$ where $\overset{\frown}{AC} \simeq \overset{\frown}{CB}$,
(b) radii $\overline{PA}$, $\overline{PB}$, and $\overline{PC}$, and
(c) $\text{int}\odot P \cap (\text{int}\angle APC \cup \text{int}\angle BPC)$.

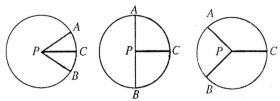

Definition	A **segment of a circle** is the union of $\overset{\frown}{ACB}$, the chord $\overline{AB}$, and the points of the interior of the circle that lie on the same side of $\overline{AB}$ as point C.

Figure 7.8

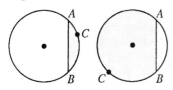

See Figure 7.8 for examples of segments. The reader should be able to show that sectors and segments satisfy the definition of a region. (See Exercise Set 7.3, No. 23.)

Consider the sequence of inscribed regular polygons in Figure 7.9 from (a) through (c). As the sides of the polygons increase in number, the perimeters approach a limiting value. This value is the **circumference** of the circle.

The following relationship has been observed to hold for all circles. As indicated in the historical note for this chapter, Archimedes was aware of its significance.

Figure 7.9

(a)

Postulate 7.5	The ratio of a circle's circumference to its diameter is the same for any circle.

The common ratio referred to in Postulate 7.5 is an irrational number denoted by the symbol π, where π is approximately equal to 3.1416. Hence, using Postulate 7.5, we easily derive the following theorem.

(b)

Theorem 7.6	**($C = 2\pi r$)** The circumference of a circle is equal to twice the product of π and the radius of the circle.

(c)

In Figure 7.9, as the sides of the polygons increase in number, the polygonal regions formed approximate the circular region more closely; thus, the limiting value of the areas of the polygonal regions equals the area of the circular region. Also notice that the lengths of the apothems of the regular polygons approach the radius of the circle. Since the area of a regular polygon is half the product of the length of its apothem and perimeter, it follows that the area of a circle is half the product of its radius and circumference. Thus, $A = rC/2$. But, by Theorem 7.6, $C = 2\pi r$. Therefore, $A = (r)(2\pi r)/2 = \pi r^2$. We have outlined the argument used in the proof of the following theorem.

Theorem 7.7	**($A = \pi r^2$)** The area of a circle is the product of π and the square of the radius of the circle.

Example 1

Find the circumference and area of a circle with radius 7 inches.

Solution

By Theorem 7.6, its circumference $C = 2\pi(7 \text{ in}) = 14\pi$ in.
By Theorem 7.7, its area $A = \pi(7 \text{ in})^2 = 49\pi$ in².

■

Example 2

Figure 7.10

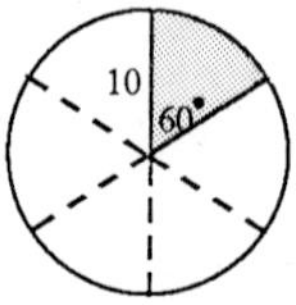

Find the area of the sector shown in Figure 7.10.

Solution

Using Theorem 7.7, we find that the area of the complete circle is 100π. It is easy to see that the area of the sector is $100\pi/6 = 50\pi/3$, since six sectors of the same area form the complete circular region.

■

In general, the area of a sector with a radius r and a central angle of measure $a°$ is $\pi a r^2/360$.

Example 3

Figure 7.11

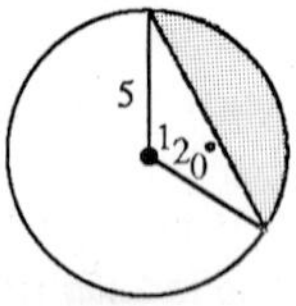

Find the area of the segment in Figure 7.11.

Figure 7.12

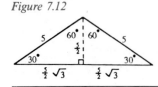

Solution

To find the area of the segment, subtract the area of the triangle from the area of the sector. The area of the sector is $\pi(120)(5)^2/360 = 25\pi/3$. The area of the triangle is $(5\sqrt{3}/2)(5/2) = 25\sqrt{3}/4$. (See Figure 7.12) Thus, the area of the segment is

$$25\pi/3 - 25\sqrt{3}/4 = (100\pi - 75\sqrt{3})/12.$$

■

EXERCISE SET 7.3

In Exercises 1-6, find the area of each of the regions, given ⊙P, radius = 2, m∠MPN = 120°.

Figure for Exercises 1-6

1. ⊙P

2. sector PMRN

3. sector PMSN

4. △MNP

5. segment MRN

6. segment MSN

In Exercises 7-12, find the perimeter of each of the regions given in Exercises 1-6.

In Exercises 13-16, find the area of each of the shaded regions.

13.

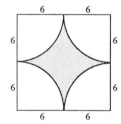

14.

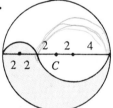

15.

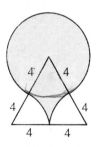

16.

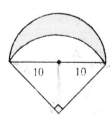

17. An *annular ring* is the union of two concentric circles with center P and radii r_1 and r_2, and the set of points Q such that $r_1 < PQ < r_2$. Find the area of the annular ring formed by circles whose radii are 5 and 10.

Figure for Exercise 18

18. A *Norman window* is formed by a semicircular region above a rectangular region as shown in the diagram. If the rectangular portion is 4 feet wide and $(4 + \pi)/2$ feet high, find the area and perimeter of the window.

19. Construct a *trefoil*. (See *multifoil* in a mathematical dictionary.) If a side of the triangle is 4, find the area and perimeter of the trefoil.

20. Construct a *quatrefoil*. (See *multifoil* in a mathematical dictionary.) If a side of the square is 3, find the area and perimeter of the quatrefoil.

21. Prove that the formula for the length L of the arc of a sector whose radius is r and whose central angle has a measure of $a°$ is $L = a\pi r/180$.

22. Prove that if two arcs have equal radii, then the ratio of their lengths equals the ratio of their measures.

23. Show that a sector and a segment of a circle satisfy the definition of region given in Section 7.1.

24. If a circular pizza 6 inches in diameter serves one person, how many circular 10-inch diameter pizzas are needed to serve 15 people?

25. If a circular pizza 6 inches in diameter serves one person, how many circular 12-inch diameter pizzas are needed to serve 20 people?

Chapter 7
Key Terms

Chapter 7
Review Exercises

Fill each blank with the word *always, sometimes,* or *never*.

1. A segment of a circle is _____ a sector.

2. The area of a triangle is _____ half the product of the lengths of two of its sides.

3. The area of a square is _____ less than the area of the circle whose diameter has the same length as the side of the square.

4. The union of two regions is _____ a region.

5. The area of an inscribed polygon is _____ greater than the area of the circumscribed circle.

True-False: If the statement is true, mark it so. If it is false, replace the underlined word to make a true statement.

6. If the lengths of adjacent sides of a rectangle and of a parallelogram with an acute angle are a and b, then the rectangle has the <u>smaller</u> area.

7. If a side of a square is a units long and a side of a second square is b units long, the ratio of the areas is <u>a/b</u>.

8. If the length of a side of a triangle is doubled and the length of the altitude to that side is tripled, the area of the new triangle is <u>five</u> times that of the original triangle.

9. The area A of a sector of a circle with radius r must satisy the inequality <u>$0 < A < \pi r^2$</u>.

10. If the radius of a circle is r and a side of a square is $r\sqrt{\pi}$ units long, the areas of the circle and square are <u>equal</u>.

Answer the following questions.

11. The length of a side of a triangle is twice the length of the altitude to that side. If the area of the triangle is 12, find the lengths of the side and altitude.

12. If a triangle is inscribed in a semicircle as shown, find the maximum area the triangle can have if $PA = 5$.

13. If $m\angle A = 30°$ and $PA = 4$ in the figure, find the area of the segment ADC.

Figure for Exercises 12–13

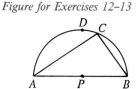

In Exercises 14–16, if the length of the shorter base of a trapezoid is 10, the base angles at the longer base are 30° and 45°, respectively, and the ratio of the area to the altitude is 11:1, find:

14. The length of the longer base.

15. The length of the altitude.

16. The area of the trapezoid.

17. If a regular hexagon is inscribed in a circle whose radius is 8, find the total area of the segments of the circle.

Proofs:

18. If the length of a side of an equilateral triangle is s, prove that the length of an altitude is $s\sqrt{3}/2$, and that the area of the triangle is $s^2\sqrt{3}/4$.

19. Prove that if the length of a side of one parallelogram equals the length of the corresponding side of another parallelogram, then the ratio of the areas equals the ratio of lengths of the altitudes to the congruent sides.

20. Prove that the area formula for a rectangle is a special case of the area formula for a trapezoid.

21. Prove that the area formula for a parallelogram is a special case of the area formula for a trapezoid.

22. Prove that the area formula for a triangle is a special case of the area formula for a trapezoid.

In Exercises 23–26, find the area of each of the shaded regions.

23.

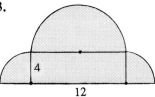

24.

25.

26.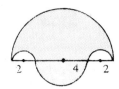

27. If △PST and △PQR are equilateral, find the area of the shaded region.

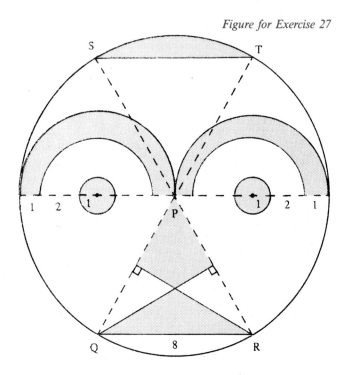

Figure for Exercise 27

Coordinate Geometry

Courtesy of Historical Pictures Service, Chicago

René Descartes (1596–1650)

Historical Note

The great French philosopher René Descartes (1596–1650) revolutionized geometry with the invention of the concept of a coordinate system, which enabled mathematicians to interpret algebraic equations geometrically through the use of curves. This system is known today as the Cartesian coordinate system, sometimes called the rectangular coordinate system. Descartes also developed the first systematic classification of curves; in fact, he tried to geometrize all of nature.

In Chapter 1, we discussed the concept of one-dimensional co-ordinates, that is, coordinates on a line. In elementary algebra, you learned about coordinates in a plane, as well as on a line. In this chapter, we relate the topic of planar coordinates to plane geometry, giving a brief review of some of the concepts from algebra.

8.1 TWO-DIMENSIONAL COORDINATE SYSTEMS

Consider a two-dimensional coordinate system in which two perpendicular coordinate lines intersect at the zero coordinate on each line. One of the lines, the **x-axis**, is placed in a horizontal position, and the other line, the **y-axis**, is placed vertically. Any point in the plane containing these lines is then assigned a pair of coordinates. The **x-coordinate**, or **abscissa**, of the point is found by passing a line through the point so that the line is perpendicular to the x-axis. The intersection of the line and the x-axis yields a point. The one-dimensional coordinate of this point is the abscissa. The **y-coordinate**, or **ordinate**, of the point is found similarly by passing a line through the point so that the line is perpendicular to the y-axis. The intersection of the line and the y-axis yields a point. The one-dimensional coordinate of this point is the ordinate. Corresponding to the original point is an **ordered pair**, in which the first member is the abscissa and the second member is the ordinate. In particular, the intersection of the two axes is the point $(0, 0)$, called the **origin**. In Figure 8.1, point A has coordinates $(4, 3)$, point B has coordinates $(-2, 4)$, point C has coordinates $(-1, -5)$, point D has coordinates $(3, -2)$, point E has coordinates $(8, 0)$, and point F has coordinates $(0, 3)$.

Using this coordinate system, we can now indicate the placement of geometric objects in a plane.

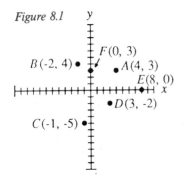

Figure 8.1

Figure 8.2

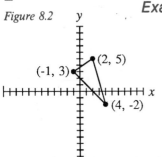

Example 1

Graph the triangle with vertices at $(2, 5)$, $(-1, 3)$, and $(4, -2)$.

Solution

After plotting the given points and joining the vertices, we get the triangle shown in Figure 8.2.

■

Example 2

Graph the quadrilateral with vertices at (7, 1), (7, 8), (−2, 3), and (−2, −4).

Figure 8.3

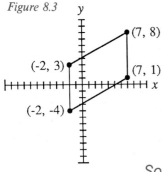

Solution

The quadrilateral is shown in Figure 8.3.

■

The quadrilateral in Figure 8.3 looks like a parallelogram. By Theorem 3.16, one way to show that it is a parallelogram is to show that opposite sides are congruent. In order to do this, we need to find the distance between two points in a plane by using the coordinates of the points. This leads us to Theorem 8.1.

Theorem 8.1	$(PQ = \sqrt{(a - c)^2 + (b - d)^2})$ The distance between the points $P(a, b)$ and $Q(c, d)$ equals the square root of the sum of the squares of $(a - c)$ and $(b - d)$.

The proof of Theorem 8.1 is left as an exercise.

■

Example 3

Find the distance between the two points $A(-1, 3)$ and $B(-4, -7)$.

Solution

$$AB = \sqrt{(-1 + 4)^2 + (3 + 7)^2} = \sqrt{9 + 100} = \sqrt{109}$$

∎

Example 4

Prove that the quadrilateral of Example 2 is a parallelogram.

Solution

The *left* side has length $\sqrt{(-2 + 2)^2 + (3 + 4)^2}$, which equals 7. The *right* side has length $\sqrt{(7 - 7)^2 + (8 - 1)^2}$, which also equals 7. The *top* side has length $\sqrt{(7 + 2)^2 + (8 - 3)^2}$, which equals $\sqrt{106}$. The *bottom* side has length $\sqrt{(7 + 2)^2 + (1 + 4)^2}$, which also equals $\sqrt{106}$. Therefore, by Theorem 3.16, the quadrilateral is a parallelogram.

∎

Example 5

Find the ordinate of the point P, which satisfies the condition that P has abscissa of -2 and lies on the circle with center at $C(-1, 4)$ and one endpoint of diameter at $A(1, 5)$.

Solution

Let the point P have coordinates $(-2, y)$. The radius of the circle is

$$AC = \sqrt{(-1 - 1)^2 + (4 - 5)^2} = \sqrt{5},$$

but $PC = AC$. Thus,

$$PC = \sqrt{(-2 + 1)^2 + (y - 4)^2} = \sqrt{5}.$$

It follows that

$$1 + (y - 4)^2 = 5,$$
$$(y - 4)^2 = 4,$$
$$y - 4 = \pm 2,$$
$$y = 4 \pm 2 = 2 \text{ or } 6.$$

∎

EXERCISE SET 8.1

Draw a graph for each exercise.

Find the distance between the two points in Exercises 1–6.

1. (6, 8) and (2, 5)

2. (4, 1) and (3, −3)

3. (0, 0) and (7, 24)

4. (−8, 11) and (13, −10)

5. (0, 8) and (−5, 0)

6. (7, −5) and (7, 8)

7. Find the perimeter of the triangle with vertices at the points (5, −8), (−3, 0), and (2, −5).

8. Show that the triangle with vertices at the points (−4, 3), (0, 6), and (2, −5) is a right triangle.

In Exercises 9–11, answer the following questions. Is the quadrilateral with the given vertices a parallelogram? Is it a rhombus? Is it a rectangle? Is it a square?

9. (2, −2), (−4, 6), (−8, 3), and (−2, −5).

10. (−4, 14), (6, −1), (1, −9), and (3, 7).

11. (4, 5), (−6, 2), (−3, −8), and (7, −5).

In Exercises 12–16, find the coordinates of the point P that satisfy the given conditions.

12. The origin is the midpoint of the line segment from P to $A(2, 5)$.

13. The x-axis is the perpendicular bisector of the line segment from P to $B(−3, 8)$.

14. The abscissa and ordinate of P are equal, and P is $10\sqrt{2}$ units from the origin. (two answers)

15. P is on the positive y-axis, and its distance from the origin is the same as the distance from $(−7, 0)$ to the origin.

16. P has an ordinate of 7, and it lies on the circle with center at $(2, 3)$ and radius of 8. (two answers)

17. Prove the formula for the distance between two points. (Hint: Use the Pythagorean Theorem.)

8.2 LINES AND LINE SEGMENTS

Figure 8.4

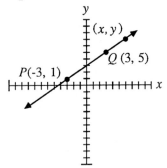

We know by Postulate 1.5 that for any two distinct points in a plane, the line containing the points is also in the plane. In Figure 8.4, we illustrate the line that passes through the points $P(-3, 1)$ and $Q(3, 5)$.

Definition

> The **slope** of a line which passes through the points $P(a, b)$ and $Q(c, d)$ equals $(b-d)/(a-c)$.

Notice in the definition above that the ratio $(b-d)/(a-c)$ equals the ratio $(d-b)/(c-a)$. We indicate how the line *tilts* by defining the slope of a line as the ratio of the vertical change to the nonzero horizontal change of any two points on the line. If the horizontal change is zero, then we say that the slope does not exist, and the line is a **vertical line**.

Example 1

Find the slope of the line in Figure 8.4.

Solution

The vertical change from P to Q is 4, found by calculating $5 - 1$. Likewise, the horizontal change is $3 - (-3) = 6$. Thus, the slope is 4/6 or 2/3. In computing the slope, we subtracted the coordinates of P from the coordinates of Q. We would obtain the same result if we subtracted the coordinates of Q from the coordinates of P, namely $(1 - 5)/(-3 - 3) = 2/3$.

■

■

Example 2

Find the equation of the line in Figure 8.4.

Solution

Given an arbitrary point (x, y) on the line in Figure 8.4, the slope from point P to this point is $(y - 1)/(x + 3)$, and since we already know from Example 1 that the slope of the line is 2/3, we have the equation of the line $(y - 1)/(x + 3) = 2/3$. Multiplying both sides of the equation by the least common denominator $3(x + 3)$ we have $3(y - 1) = 2(x + 3)$. Distributing to eliminate the parentheses, we obtain $3y - 3 = 2x + 6$, which can be rewritten as $2x - 3y + 9 = 0$.

■

Definition

> The **general form** of the equation of a line is $ax + by + c = 0$, where a, b, and c are real numbers (integers if possible).

Instead of using the general form of a line, it is often helpful to use the slope-intercept form of a line.

Definition

> The **slope-intercept form** of a line is $y = mx + b$, where m is the slope of the line and b is the y-intercept.

In the above definition, when we refer to the y-intercept, we mean the ordinate of the point of intersection of the line and the y-axis.

■

Example 3

Find the slope and the y-intercept of the line in Figure 8.4.

Solution

In Example 2, if we solve the equation for y, we obtain $y = (2/3)x + 3$; thus, the slope is 2/3 and the y-intercept is 3.

■

In Chapter 1, we learned that a line contains infinitely many points. In general, we can find other points on a line by substituting arbitrary x-values in the equation of a line and solving for the corresponding y-values. For example, if we substitute the value $x = 6$ in the equation $y = (2/3)x + 3$, we find that $y = 7$. Thus, another point on the line is $(6, 7)$. Substituting $x = -5$, we find that $y = -1/3$. Thus, $(-5, -1/3)$ is a point on the line.

To designate a line segment, we restrict the values of x. We define an **interval [a, b]** of x-values to be the union of a, b, and all numbers between a and b. If a line is restricted by allowing only x-values in an interval, then we obtain a line segment.

Example 4

Graph the equation of the line segment found by restricting the x-values of the line $y = 2x - 5$, to the interval $[0, 4]$.

Figure 8.5

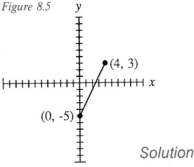

Solution

The graph of the line segment is shown in Figure 8.5. The endpoints of the line segment can be found by using the values $x = 0$ and $x = 4$, which bound the interval. Thus, one endpoint has coordinates $(0, -5)$, and the other has coordinates $(4, 3)$.

We could have defined a line segment by restricting the y-values to an interval, a restriction that is sometimes necessary. Consider the general form of a line $ax + by + c = 0$, where $a = 1$, $b = 0$ and $c = -2$. This simplifies to the equation $x = 2$. There are infinitely many ordered pairs in which the abscissa is 2. For example, $(2, -5)$, $(2, 500)$, and $(2, 0)$ are three points on the line. Clearly, we cannot consider a restriction of x-values in order to yield a line segment. However, we could restrict the y-values to the interval $[-2, 5]$.

Example 5

Draw the graph of the line segment found by restricting the y-values of the line $x = 2$ to the interval $[-2, 5]$.

Figure 8.6

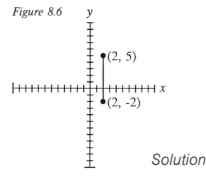

Solution

The graph is shown in Figure 8.6.

■

If c is a constant, a line of the form $x = c$ is a vertical line and a line of the form $y = c$ is a horizontal line. The **midpoint** of a line segment is found by averaging the x-coordinates and averaging the y-coordinates of the endpoints.

Example 6

Find the midpoint of the line segment in Figure 8.5 and the line segment in Figure 8.6.

Solution

The line segment in Figure 8.5 has a midpoint at $(2, -1)$ since $(0 + 4)/2 = 2$ and $(-5 + 3)/2 = -1$. The line segment in Figure 8.6 has endpoints at $(2, -2)$ and $(2, 5)$. Thus, its midpoint is at $(2, 3/2)$ since $(2 + 2)/2 = 2$ and $(-2 + 5)/2 = 3/2$.

■

Two lines are parallel if they have the same slope. Two lines are perpendicular if the product of their slopes equals -1. In the event that one of the lines we are considering is a vertical line, then as previously mentioned, the slope does not exist. In this case, it is clear that a line parallel to a vertical line is another vertical line, and a line perpendicular to a vertical line is a horizontal line.

■

Example 7

Consider the following pairs of lines and determine whether they are parallel or perpendicular lines.

(a) $y = -2x + 3$ and $y = -2x - 5$
(b) $y = (3/4)x + 7$ and $y = (-4/3)x + 9$
(c) $x = 4$ and $x = 13$
(d) $x = -5$ and $y = 3$

Solution

(a) Parallel, since both lines have a slope of -2.
(b) Perpendicular, since $3/4$ times $-4/3$ equals -1.
(c) Parallel, since both lines are vertical.
(d) Perpendicular, since $x = -5$ is a vertical line and $y = 3$ is a horizontal line.

■

EXERCISE SET 8.2

Draw a graph for each exercise. Find the slope of the lines through the given points in Exercises 1-6.

1. $(2, 5)$ and $(7, 10)$

2. $(-3, 4)$ and $(6, -2)$

3. $(8, 0)$ and $(9, 2)$

4. $(3, -2)$ and $(8, -2)$

5. $(4, 7)$ and $(0, 0)$

6. $(5, -3)$ and $(5, 7)$

7. Find the midpoint of the line segment whose endpoints are given in Exercises 1-6.

8. Find the center of the circle having a diameter whose endpoints are $(-3, 5)$ and $(4, -2)$.

9. If $(2, 4)$ is the center of a circle, and $(-5, -3)$ is an endpoint of a diameter, find the other endpoint of the diameter.

10. Write the equation of the line through each pair of points in Exercises 1-6. Write answers in general form.

Find the slope and *y*-intercept of the line represented by each of the following equations in Exercises 11-16.

11. $2y = 3 - 4x$

12. $5x - 7y - 8 = 0$

13. $y + 2 = 3(x - 5)$

14. $y = -3$

15. $x = 7y - 6$

16. $2 - \pi x = 9y$

17. Write the equation of the line through the point $(0, 8)$ and having slope of $-1/3$.

Write the equation of the line that passes through the point $(-7, 2)$ and has the following properties:

18. It has a slope of zero.

19. It is parallel to the line $2x - 5y = 10$.

20. It is perpendicular to the line $3x + 8y = 7$.

21. It is vertical.

22. It is horizontal.

Draw the graphs of each of the following line segments:

23. $y = 4x - 8$, for the *x*-interval $[0, 3]$.

24. $x + 3y = 6$, for the *y*-interval $[-1, 2]$.

25. $4x - 3y + 12 = 0$, for the *x*-interval $[-3, 3]$.

26. $y = -1$, for the *x*-interval $[0, 5]$.

27. $x = 4$, for the *y*-interval $[-6, -2]$.

28. Write the equations of the altitudes of the triangle whose vertices are $A(-2, 8)$, $B(1, -5)$, and $C(8, 3)$. Also, find the point at which the altitudes are concurrent.

29. Write the equations of the medians of $\triangle ABC$ in Exercise 28, and find the point at which the medians are concurrent.

30. Write the restricted equations to represent the sides of $\triangle ABC$ in Exercise 28.

8.3 POLYGONS

To represent a line segment, we find the equation of the line containing the line segment, and then we restrict the x-values and/or y-values appropriately. To represent a polygon algebraically, we must find the restricted equations of the line segments corresponding to the sides of the polygon. A triangle is described by listing three restricted line equations. A pentagon is described by listing five restricted line equations. An n-gon is described by listing n restricted line equations.

Describing geometric objects in this way is burdensome because in the past, if we were given vertices of a polygon, we merely took a straightedge and joined the vertices with the appropriate line segments. We did not seem to need their equations. However, even though the use of geometrical constructions is perfect in theory, in practice, constructions yield approximate results, and if we need more accuracy, we should work algebraically.

Also, if we are interested in using the capabilities of a computer to draw our diagrams, we need to be able to describe the correct set of points to be graphed. Line segments are commonly used in computer graphics.

■

Example 1

Consider the triangle in Figure 8.7 with vertices at $A(-2, 3)$, $B(4, 7)$, and $C(3, -5)$. Find the equations of line segments $\overline{AB}$, $\overline{AC}$, and $\overline{BC}$.

Solution

Figure 8.7

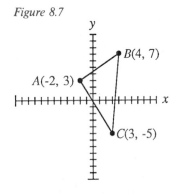

The line $\overleftrightarrow{AB}$ has slope $(7 - 3)/(4 + 2) = 2/3$. The line $\overleftrightarrow{AC}$ has slope $(-5 - 3)/(3 + 2) = -8/5$. The line $\overleftrightarrow{BC}$ has slope $(-5 - 7)/(3 - 4) = 12$. Thus, the equations of the three lines respectively, are

$$(y - 3)/(x + 2) = 2/3,$$
$$(y - 3)/(x + 2) = -8/5$$
$$(y - 7)/(x - 4) = 12.$$

Multiplying each equation by the least common denominator and writing the equations in general form, we obtain the three equations

$$2x - 3y + 13 = 0,$$
$$8x + 5y + 1 = 0,$$
$$12x - y - 41 = 0.$$

(We used point A to determine the equations of the first two lines, and point B for the third line. We could have used the other endpoints instead. In any case, the resulting *simplified* equations will be identical.)

Finally, in order to determine the restrictions on the lines that will form the sides of the polygons, we use the x-values of the vertices, which are the x-values of the endpoints of the line segments we wish to have. We summarize the final results:

Line segment AB: $2x - 3y + 13 = 0$, $[-2, 4]$.
Line segment AC: $8x + 5y + 1 = 0$, $[-2, 3]$.
Line segment BC: $12x - y - 41 = 0$, $[3, 4]$.

∎

The general custom is to indicate the restriction of values by using the interval of x-values. However, if the line segment happens to be vertical, then we must restrict the y-values, as previously mentioned.

We now give an example of a proof which uses the methods of coordinate geometry. Although proofs can be done by the traditional methods used earlier in this text, some individuals believe that coordinate algebraic proofs are easier to understand and implement. However, in doing proofs by this method, we must be careful not to use circular reasoning. That is, we must not use information that we have obtained from the theorem which we proved in the traditional manner, while proving that theorem using coordinate geometry.

∎

Example 2

Prove that the diagonals of a parallelogram bisect each other.

Solution

Figure 8.8

Place the parallelogram $ABCD$ on a coordinate system as shown in Figure 8.8. The y-coordinates of points B and C are equal, since by Theorem 3.17, two parallel lines are everywhere equidistant, and $\overline{BC}$ is parallel to the x-axis.

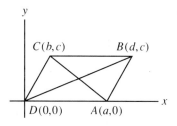

Given: $ABCD$ is a parallelogram

Prove: $\overline{AC}$ and $\overline{BD}$ intersect at their midpoints

Proof: Since $ABCD$ is a parallelogram, the opposite sides are parallel. Thus, the opposite sides have the same slope. It follows that $(c - 0)/(b - 0) = (c - 0)/(d - a)$ so that $c/b = c/(d - a)$, which simplifies to $b = d - a$, or $d = a + b$. Thus, the midpoint of $\overline{BD}$ is $((a + b)/2, c/2)$, which is also the midpoint of $\overline{AC}$. Therefore, $\overline{AC}$ and $\overline{BD}$ intersect at their midpoints.

∎

EXERCISE SET 8.3

Draw a graph and write the restricted equations for the sides of each polygon in Exercises 1-3.

1. Vertices $(3, 8)$, $(-6, 2)$, $(-1, -7)$

2. Vertices $(8, 5)$, $(2, 4)$, $(-2, 1)$, $(3, 3)$.

3. Vertices $(-6, 0)$, $(0, 10)$, $(6, 5)$, $(0, 5)$, $(-3, 2)$.

Draw a graph for Exercises 4-6.

4. The sides of a quadrilateral lie on lines $x = 0$, $y = 0$, $2x - 4y = 8$, and $5x + 3y = 30$. What are the restrictions for the sides?

5. One side of a square is given by $2x - 5y - 13 = 0$, x-interval $[-1, 4]$. Find the restricted equations of the other three sides of the square. (two answers)

6. Four sides of an octagon are parallel to the axes and are represented by $x = 5$, $[-2, 2]$; $y = 5$, $[-2, 2]$; $x = -5$, $[-2, 2]$; and $y = -5$, $[-2, 2]$. Find the restricted equations for the slanting sides joining the given sides. Is it a regular octagon? Why?

7. Graph the following polygons on the same coordinate system.
   Quadrilateral $(1, -4)$, $(2, -3)$, $(-2, -3)$, $(-1, -4)$
   Quadrilateral $(0, 12)$, $(-8, 8)$, $(0, -10)$, $(8, 8)$
   Triangle $(-2, 8)$, $(-3, 6)$, $(-4, 8)$
   Triangle $(2, 8)$, $(3, 6)$, $(4, 8)$
   Triangle $(0, 5)$, $(-1, 1)$, $(1, 1)$

8. Find the restricted equations for the *face* in Chapter 7 Review Exercises, No. 27.

Prove Exercises 9–12 using coordinate geometry.

9. Prove Theorem 3.22: If a segment joins the midpoints of two sides of a triangle, then it is parallel to the third side and has half the length of the third side.

10. Prove Theorem 3.24: The median of a trapezoid is parallel to both bases and its length is equal to half the sum of the lengths of the bases.

11. Prove that the three medians of a triangle intersect in a point two-thirds of the distance from each vertex to the midpoint of its opposite side. This point is the **centroid** of the triangle. (See the Chapter 6 Review Exercises, No. 46.)

12. Prove that the three altitudes of a triangle intersect in a point. This point is the **orthocenter** of the triangle. (See the Chapter 6 Review Exercises, No. 46.)

8.4 CIRCLES

In Section 8.1, we introduced the formula for finding the distance between two points. We can use this formula to find the equation of a circle. We have defined a circle to be the set of all points in a plane at a given distance (the radius) from a given point (the center). Let the radius of a circle be r and its center have coordinates $C(h, k)$. Given an arbitrary point (x, y) on the circle, the distance between this point and the center of the circle is r. Thus by Theorem 8.1, we have

$$\sqrt{(x - h)^2 + (y - k)^2} = r.$$

Squaring both sides of this equation yields the equation of the circle

$$\boxed{(x - h)^2 + (y - k)^2 = r^2}$$

called the **standard form** of the equation of a circle.

If we take the standard form of the equation and square the parenthetical terms we get

$$x^2 - 2hx + h^2 + y^2 - 2ky + k^2 = r^2.$$

Rearranging terms, we can write this as

$$x^2 + y^2 - 2hx - 2ky + h^2 + k^2 - r^2 = 0.$$

Setting $a = -2h$, $b = -2k$, and $c = h^2 + k^2 - r^2$ yields the **general form** of the equation of a circle

$$\boxed{x^2 + y^2 + ax + by + c = 0.}$$

When graphing a circle on a coordinate system, we again use a compass, rather than plotting separate points.

Figure 8.9

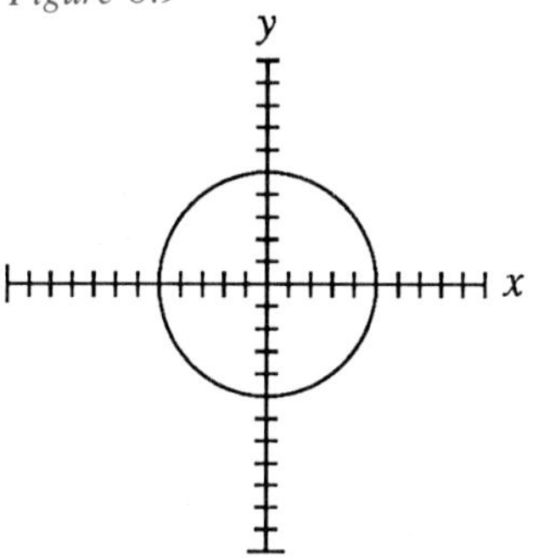

Example 1

Find the equation of the circle with radius 5 and center at $(0, 0)$. Also, graph the circle.

Solution

The circle has equation $x^2 + y^2 = 25$, which can also be written in the general form $x^2 + y^2 - 25 = 0$. See Figure 8.9 for the graph of the circle.

■

Example 2

Consider a circle with radius 2 and center at $(-1, 3)$. Find its equation and graph it.

Solution

Figure 8.10

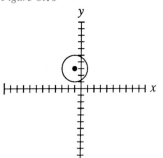

The standard form of the equation of this circle is

$$(x + 1)^2 + (y - 3)^2 = 4.$$

Squaring the parenthetical terms we get

$$x^2 + 2x + 1 + y^2 - 6y + 9 = 4,$$

which can be written in the general form

$$x^2 + y^2 + 2x - 6y + 6 = 0.$$

See Figure 8.10 for its graph.

■

A clear advantage of having the equation of a circle written in standard form is that it is easy to find its center and radius. However, the general form of the equation is simpler, since like terms have been combined. Given the equation of a circle in general form, we can find its center and radius by converting the equation to standard form. This requires the *completing-the-square* process learned in algebra.

■

Example 3

Find the center and radius of the circle with equation

$$x^2 + y^2 - 6x + 2y - 7 = 0.$$

Solution

We rearrange the terms and write this equation as

$$x^2 - 6x + y^2 + 2y = 7.$$

Since half of -6 (the coefficient of x) is -3, we add $(-3)^2$ to both sides of the equation, and since half of 2 (the coefficient of y) is 1, we add $(1)^2$ to both sides of the equation. This yields

$$x^2 - 6x + 9 + y^2 + 2y + 1 = 17.$$

Factoring, we get

$$(x - 3)^2 + (y + 1)^2 = 17,$$

thus, the radius of the circle is $\sqrt{17}$ and its center is at $(3, -1)$.

■

EXERCISE SET 8.4

Graph each of the circles in Exercises 1–4, and write its equation in standard form.

1. Center $(0, 0)$, radius 3.

2. Center $(1, 3)$, radius 5.

3. Center $(-4, -7)$, radius 4.

4. Center $(4, -2)$, radius $2\sqrt{5}$.

Write the equation of the circle in general form for each of Exercises 5–10.

5. Center $(4, -8)$, radius 13.

6. Endpoints of a diameter are $(7, 3)$ and $(-3, -5)$.

7. Center $(-3, 6)$ and passing through $(5, 1)$.

8. Center $(4, -6)$ and tangent to the x-axis.

9. Center $(4, -6)$ and tangent to the y-axis.

10. Passing through the points $(0, 0)$, $(0, -5)$, and $(8, 0)$.

Find the center and radius of each of the circles in Exercises 11–14 by first rewriting the equation in standard form.

11. $x^2 + y^2 - 4x - 10y + 20 = 0$.

12. $x^2 + y^2 - 16y + 39 = 0$.

13. $x^2 + y^2 + 14x - 25y + 57 = 0$.

14. $9x^2 + 9y^2 - 12x + 60y + 55 = 0$.

15. Write the equation of the concentric circles having center at $(-3, 4)$ and respective radii of 3, 4, 5, and 6. Write answers in general form.

16. Graph circle $x^2 + y^2 - 8x - 2y - 23 = 0$ and line $2x - 2y - 1 = 0$ on the same coordinate system. Estimate the coordinates of the two points of intersection. Can these coordinates be found exactly? How?

8.5 TRANSFORMATIONS

In this section we analyze some relationships between sets of points. The concepts we discuss do not necessarily require the use of coordinate geometry, but we find it convenient to illustrate the concepts through this medium.

■

Definition

> A **transformation** $\mathcal{T}$ is a one-to-one correspondence between two sets of points $\mathbb{R}$ and $\mathbb{S}$.

We write $\mathcal{T} : \mathbb{R} \leftrightarrow \mathbb{S}$ to represent the transformation $\mathcal{T}$ between set $\mathbb{R}$ and set $\mathbb{S}$. Some treatments of this topic refer to a transformation *from* one set to another and refer to the transformation in the other direction as an inverse transformation. However, for the purposes of this text, we prefer to limit the discussion to the concepts of one-to-one correspondence. This general definition refers to a one-to-one correspondence between sets, but has no other restrictions. An example of a transformation that has further restrictions is given in the following definition.

■

Example 1

Determine whether the following are transformations between set $\mathbb{R}$ and set $\mathbb{S}$.

(a) $\mathbb{R} = \{x : x = -3, -2, -1, 0, 1, 2, 3\}$, $\mathbb{S} = \{z : z = -6, -4, -2, 0, 2, 4, 6\}$, where $\mathcal{T} : \mathbb{R} \leftrightarrow \mathbb{S}$ is defined by $z = 2x$.

(b) $\mathbb{R} = \{y : y = -4, 0, 4\}$, $\mathbb{S} = \{p : p = 0, 16\}$, where $\mathcal{T} : \mathbb{R} \leftrightarrow \mathbb{S}$ is defined by $p = y^2$.

Solution

(a) $\mathcal{T}$ is a transformation since there is a one-to-one correspondence between $\mathbb{R}$ and $\mathbb{S}$. The correspondence is as follows:

$$-3 \leftrightarrow -6 \quad -2 \leftrightarrow -4 \quad -1 \leftrightarrow -2 \quad 0 \leftrightarrow 0 \quad 1 \leftrightarrow 2 \quad 2 \leftrightarrow 4 \quad 3 \leftrightarrow 6$$

(b) $\mathcal{T}$ is not a transformation, since $p = y^2$ does not define a one-to-one correspondence between $\mathbb{R}$ and $\mathbb{S}$. In particular, 16 in $\mathbb{S}$ corresponds to two elements of $\mathbb{R}$, namely -4 and 4. Notice also that the number of elements in $\mathbb{R}$ does not equal the number of elements in $\mathbb{S}$, so that there is no possibility of defining a one-to-one correspondence between the sets.

■

Definition	An **isometry** is a transformation that preserves distance.

This definition indicates that a transformation $\mathcal{T} : \mathbb{R} \leftrightarrow \mathbb{S}$ is an isometry if the distance between any two points R_1 and R_2 in $\mathbb{R}$ is the same as the distance between their corresponding points S_1 and S_2 in $\mathbb{S}$, that is, $R_1R_2 = S_1S_2$, where R_1 corresponds to S_1 and R_2 corresponds to S_2. We introduce three categories of isometries, namely translations, rotations, and reflections. These are defined later in this section.

■

Example 2

Determine whether the following is an isometry between set $\mathbb{R}$ and set $\mathbb{S}$. $\mathbb{R} = \{(x, y): y = 2x - 9\}$, $\mathbb{S} = \{(x, y): y = 2x + 6\}$, where $\mathcal{T} : \mathbb{R} \leftrightarrow \mathbb{S}$ is defined as follows: Match each point in $\mathbb{R}$ to the point in $\mathbb{S}$ obtained by keeping the same x-coordinate and adding 15 to the y-coordinate of the point in $\mathbb{R}$.

Solution

The relation $\mathcal{T}$ is a transformation since it is a one-to-one correspondence between $\mathbb{R}$ and $\mathbb{S}$. This follows from the observation that a point in $\mathbb{R}$ is of the form $(x, 2x - 9)$, and adding 15 to the y-coordinate yields the point $(x, 2x + 6)$, which is a point in $\mathbb{S}$. Let R_1 and R_2 be two arbitrary points in $\mathbb{R}$ corresponding to points S_1 and S_2 in $\mathbb{S}$. We will show that $(R_1R_2)^2 = (S_1S_2)^2$. Since distances are positive, this will imply that $R_1R_2 = S_1S_2$. By Theorem 8.1,

$$
\begin{aligned}
(R_1R_2)^2 &= (x_1 - x_2)^2 + (y_1 - y_2)^2 \\
&= (x_1 - x_2)^2 + [(2x_1 - 9) - (2x_2 - 9)]^2 \\
&= (x_1 - x_2)^2 + [2x_1 - 2x_2]^2
\end{aligned}
$$

and

$$
\begin{aligned}
(S_1S_2)^2 &= (x_1 - x_2)^2 + [(y_1 + 15) - (y_2 + 15)]^2 \\
&= (x_1 - x_2)^2 + [(2x_1 + 6 + 15) - (2x_2 + 6 + 15)]^2 \\
&= (x_1 - x_2)^2 + [2x_1 - 2x_2]^2
\end{aligned}
$$

Thus,

$$
(R_1R_2)^2 = (S_1S_2)^2
$$

yielding the result that $\mathcal{T}$ is an isometry.

■

Theorem 8.2	An isometry preserves angle measure.

Proof: Let $\mathcal{T} : \angle ABC \leftrightarrow \angle DEF$ be an isometry. Without loss of generality, we can   assume that $A \leftrightarrow D, B \leftrightarrow E$, and $C \leftrightarrow F$. (This follows since any point on $\overrightarrow{ED}$ and any point on $\overrightarrow{EF}$, other than the vertex E, can be used to define $\angle DEF$, so we choose the points that are most convenient to us and name them D and F, namely the points that satisfy $AB = DE$ and $BC = EF$.) Since an isometry preserves distance, $AC = DF$. Drawing auxiliary line segments $\overline{AC}$ and $\overline{DF}$, we have $\triangle ABC \simeq \triangle DEF$ by SSS = SSS. Since corresponding parts of congruent triangles are congruent, $\angle ABC \simeq \angle DEF$. Thus, m$\angle ABC$ = m$\angle DEF$, yielding the desired result that $\mathcal{T}$ preserves angle measure.

Definition	A **translation** is a transformation $\mathcal{T} : \mathbb{R} \leftrightarrow \mathbb{S}$ in which every point in $\mathbb{R}$ with coordinates (x, y) corresponds to a point in $\mathbb{S}$ with coordinates $(x + a, y + b)$.

Figure 8.11

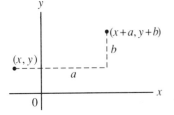

In Figure 8.11, we graph a point with coordinates (x, y) and the corresponding point with coordinates $(x + a, y + b)$. The definition of a translation between sets $\mathbb{R}$ and $\mathbb{S}$ requires that we have a correspondence such that each point in $\mathbb{R}$ corresponds to a point in $\mathbb{S}$.

Example 3

Determine whether the set $\mathbb{S} = \{(x, y): y = x + 10\}$ is a translation of the set $\mathbb{R} = \{(x, y): y = x + 3\}$.

Solution

Points in $\mathbb{S}$ are determined by the equation $y = x + 10$. Thus, if $\mathcal{T}$ is a translation, a point in $\mathbb{S}$ with coordinates $(x + a, y + b)$ must satisfy the equation $y + b = x + a + 10$. Since points in $\mathbb{R}$ satisfy the equation $y = x + 3$, by substitution we get $x + 3 + b = x + a + 10$. Solving this equation for b, we get $b = a + 7$, which has infinitely many solutions for a and b. It follows that infinitely many translations exist between $\mathbb{R}$ and $\mathbb{S}$. For example:

if $a = 5$, then $b = 12$, so that $(x, y) \leftrightarrow (x + 5, y + 12)$;

if $a = 0$, then $b = 7$, so that $(x, y) \leftrightarrow (x, y + 7)$; or

if $a = -4$, then $b = 3$, so that $(x, y) \leftrightarrow (x - 4, y + 3)$.

∎

In Example 3, we were able to show that infinitely many translations exist between the given line and the given parallel line. (Notice that the slopes of the two lines are equal.) This is not a coincidence. Translating a line results either in a parallel line or in the same line.

In Section 8.4, we discussed the standard form of a circle $(x - h)^2 + (y - k)^2 = r^2$ with center $C(h, k)$ and radius r. Notice that this is a translation of the circle $x^2 + y^2 = r^2$ with center $C(0, 0)$ and radius r. The correspondence is $(x, y) \leftrightarrow (x - h, y - k)$.

Theorem 8.3	Translations are isometries.

Proof: Let $\mathcal{T} : \mathbb{R} \leftrightarrow \mathbb{S}$ be a translation, for which each point in $\mathbb{R}$ with coordinates (x, y) corresponds to a point in $\mathbb{S}$ with coordinates $(x + a, y + b)$. Let R_1 and R_2 be two arbitrary points in $\mathbb{R}$ corresponding to points S_1 and S_2 in $\mathbb{S}$. By Theorem 8.1,

$$(R_1R_2)^2 = (x_1 - x_2)^2 + (y_1 - y_2)^2$$

and

$$
\begin{aligned}
(S_1S_2)^2 &= [(x_1 + a) - (x_2 + a)]^2 + [(y_1 + b) - (y_2 + b)]^2 \\
&= [x_1 + a - x_2 - a]^2 + [y_1 + b - y_2 - b]^2 \\
&= (x_1 - x_2)^2 + (y_1 - y_2)^2
\end{aligned}
$$

Thus,

$$(R_1R_2)^2 = (S_1S_2)^2,$$

yielding the result that $\mathcal{T}$ is an isometry.

Definition

> A **rotation** is a transformation $\mathcal{T} : \mathbb{R} \leftrightarrow \mathbb{S}$ in which every point in $\mathbb{R}$ with coordinates (x, y) corresponds to a point $\mathbb{S}$ with coordinates $(ax - by, ay + bx)$, where $a^2 + b^2 = 1$.

Figure 8.12

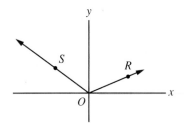

This formal definition of a rotation is not very intuitive. Informally, we think of rotation as the result of *spinning* a set about a point, such as spinning a propeller of an airplane about its center. We can also think of a rotation as a change in the position of a ray with respect to the positive x-axis. In this respect, consider the ray with the endpoint at the origin and containing a point R in $\mathbb{R}$ and another ray with endpoint at the origin and containing a point S in $\mathbb{S}$. (See Figure 8.12.) We can spin the first ray about the origin, either clockwise or counter-clockwise, until it coincides with the second ray. The point R has been rotated to point S if point R now coincides with point S. (This is guaranteed in the definition by the condition $a^2 + b^2 = 1$.) The angle formed by the two rays is called the *angle of rotation*.

Example 4

Determine whether the set $\mathbb{S} = \{(x, y): y = 2x\}$ is a rotation of the set $\mathbb{R} = \{(x, y): y = -3x\}$.

Solution

Points in $\mathbb{S}$ are determined by the equation $y = 2x$. Thus, if $\mathcal{T}$ is a rotation, a point in $\mathbb{S}$ with coordinates $(ax - by, ay + bx)$ must satisfy the equation $ay + bx = 2(ax - by)$. Since points in $\mathbb{R}$ satisfy the equation $y = -3x$, by substitution we get $-3ax + bx = 2(ax + 3bx)$. Thus, $-3ax + bx = 2ax + 6bx$ yields $-5ax = 5bx$, which simplifies to $a = -b$. This result together with $a^2 + b^2 = 1$ implies that one solution is $a = -\sqrt{2}/2, b = \sqrt{2}/2$, so that $(x, y) \leftrightarrow (-x\sqrt{2}/2 - y\sqrt{2}/2, -y\sqrt{2}/2 + x\sqrt{2}/2)$. It follows that a rotation exists between $\mathbb{R}$ and $\mathbb{S}$. For example: $(1, -3) \leftrightarrow (\sqrt{2}, 2\sqrt{2})$.

Theorem 8.4

> Rotations are isometries.

Proof: Let $\mathcal{T} : \mathbb{R} \leftrightarrow \mathbb{S}$ be a rotation, for which each point in $\mathbb{R}$ with coordinates (x, y) corresponds to a point in $\mathbb{S}$ with coordinates $(ax - by, ay + bx)$, where $a^2 + b^2 = 1$. Let R_1 and R_2 be two arbitrary points in $\mathbb{R}$ corresponding to points S_1 and S_2 in $\mathbb{S}$. By Theorem 8.1,

$$(R_1 R_2)^2 = (x_1 - x_2)^2 + (y_1 - y_2)^2$$

and

$$
\begin{aligned}
(S_1 S_2)^2 &= [(ax_1 - by_1) - (ax_2 - by_2)]^2 \\
&\quad + [(ay_1 + bx_1) - (ay_2 + bx_2)]^2 \\
&= [ax_1 - by_1 - ax_2 + by_2]^2 \\
&\quad + [ay_1 + bx_1 - ay_2 - bx_2]^2 \\
&= [a(x_1 - x_2) - b(y_1 - y_2)]^2 \\
&\quad + [b(x_1 - x_2) + a(y_1 - y_2)]^2 \\
&= a^2(x_1 - x_2)^2 - 2ab(x_1 - x_2)(y_1 - y_2) \\
&\quad + b^2(y_1 - y_2)^2 + b^2(x_1 - x_2)^2 \\
&\quad + 2ab(x_1 - x_2)(y_1 - y_2) + a^2(y_1 - y_2)^2 \\
&= a^2(x_1 - x_2)^2 + b^2(y_1 - y_2)^2 \\
&\quad + b^2(x_1 - x_2)^2 + a^2(y_1 - y_2)^2 \\
&= (a^2 + b^2)(x_1 - x_2)^2 + (a^2 + b^2)(y_1 - y_2)^2 \\
&= (x_1 - x_2)^2 + (y_1 - y_2)^2
\end{aligned}
$$

since $a^2 + b^2 = 1$. Thus,

$$(R_1 R_2)^2 = (S_1 S_2)^2,$$

yielding the result that $\mathcal{T}$ is an isometry.

Definition

> Distinct points P and Q are **symmetric with respect to line k** if P is not on k and k is the perpendicular bisector of $\overline{PQ}$. If P is on k, then P is symmetric to itself with respect to k.

Definition

> **A reflection with respect to line k** is a transformation $\mathcal{T} : \mathbb{R} \leftrightarrow \mathbb{S}$ in which every point R_1 in $\mathbb{R}$ corresponds to a point S_1 in $\mathbb{S}$, such that R_1 is symmetric to S_1 with respect to k.

Example 5

Determine whether the set $S = \{(x, y): (x - 3)^2 + (y + 1)^2 = 4\}$ is a reflection of the set $R = \{(x, y): (x + 1)^2 + (y - 3)^2 = 4\}$ with respect to the line $y = x$.

Figure 8.13

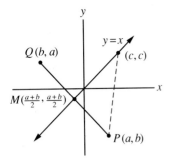

Solution

Let $P(a, b)$ be an arbitrary point in S. Then $(a - 3)^2 + (b + 1)^2 = 4$, which implies that $(b + 1)^2 + (a - 3)^2 = 4$, so that $Q(b, a)$ is in R. Using the results of Section 8.2, we find that the midpoint M of $\overline{PQ}$ is $((a + b)/2, (a + b)/2)$, so that M lies on $y = x$. This shows that $y = x$ is the bisector of $\overline{PQ}$. We now need to show that $\overline{PQ}$ is perpendicular to $y = x$. Consider the point $U(c, c)$ such that $C \neq (a + b)/2$. (See Figure 8.13.)

Point U is clearly on $y = x$ and $U \neq M$. We have

$$(PM)^2 + (UM)^2 = \left(a - \frac{a + b}{2}\right)^2 + \left(b - \frac{a + b}{2}\right)^2$$

$$+ \left(c - \frac{a + b}{2}\right)^2 + \left(c - \frac{a + b}{2}\right)^2$$

$$= \left(\frac{a - b}{2}\right)^2 + \left(\frac{b - a}{2}\right)^2 + 2\left(\frac{2c - a - b}{2}\right)^2$$

$$= \frac{(a - b)^2}{2} + \frac{(2c - a - b)^2}{2}$$

$$= \frac{1}{2}(a^2 - 2ab + b^2 + 4c^2 - 4ac + a^2 - 4bc + b^2 + 2ab)$$

$$= \frac{1}{2}(2a^2 - 4ac + 2c^2 + 2b^2 - 4bc + 2c^2)$$

$$= (a^2 - 2ac + c^2) + (b^2 - 2bc + c^2)$$

$$= (a - c)^2 + (b - c)^2$$

$$= (PU)^2.$$

By Theorem 5.7, $\triangle PUM$ is a right triangle with $m\angle PMU = 90°$. Thus, $\overline{PQ}$ is perpendicular to $y = x$. Therefore, a reflection exists between R and S with respect to $y = x$.

Figure 8.14

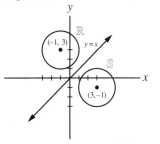

The graph in Figure 8.14 shows the sets ℝ and 𝕊 described in Example 5. Compare this graph to the illustration presented in Figure 8.13. Remember that each point P in 𝕊 corresponds to a point Q in ℝ.

In Figure 8.15 we illustrate a reflection between two regions ℝ and 𝕊.

The proof of the next theorem is left as an exercise.

Figure 8.15

Theorem 8.5	Reflections are isometries.

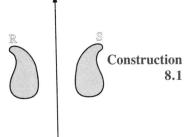

Construction 8.1	Construct the reflection of a point P with respect to a line k.

In this construction, we will assume that P is not on line k, because otherwise no construction is necessary; that is, there is a reflection $\mathcal{T}$: $P \leftrightarrow P$. Therefore, we need to find a point Q such that $\mathcal{T} : P \leftrightarrow Q$.

Given: line k, point P

Construct: Q such that $k \perp \overline{PQ}$ and k bisects $\overline{PQ}$

Step 1. Using Construction 1.6, construct the perpendicular to line k from point P. Let $M = k \cap \overline{PQ}$.

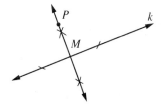

Step 2. Use a compass to draw circle M with radius PM. Circle M intersects k in two points, one of which is P. Label the other point Q. Then Q is the required point.

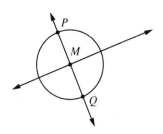

EXERCISE SET 8.5

In Exercises 1–8, determine whether the given conditions yield trans-formations between $\mathbb{R}$ and $\mathbb{S}$.

1. $\mathbb{R} = \{x\colon x = -10, -5, 0, 3, 4\}$, $\mathbb{S} = \{y\colon y = -30, -15, 0, 9, 12\}$, where $\mathscr{T}\colon \mathbb{R} \leftrightarrow \mathbb{S}$ is defined by $y = 3x$.

2. $\mathbb{R} = \{x\colon x = -16, -2, 9, 10\}$, $\mathbb{S} = \{y\colon y = -64, -8, 36, 40\}$, where $\mathscr{T}\colon \mathbb{R} \leftrightarrow \mathbb{S}$ is defined by $y = 4x$.

3. $\mathbb{R} = \{x\colon x = -5, 3, 4\}$, $\mathbb{S} = \{z\colon z = -5/2, 3/2, 2\}$, where $\mathscr{T}\colon \mathbb{R} \leftrightarrow \mathbb{S}$ is defined by $z = (1/2)x$.

4. $\mathbb{R} = \{x\colon x = -16, -2, 10\}$, $\mathbb{S} = \{z\colon z = -16/3, -2/3, 10/3\}$, where $\mathscr{T}\colon \mathbb{R} \leftrightarrow \mathbb{S}$ is defined by $x = (1/3)z$.

5. $\mathbb{R} = \{p\colon p \text{ is an integer}\}$, $\mathbb{S} = \{q\colon q \text{ is an integer}\}$, where $\mathscr{T}\colon \mathbb{R} \leftrightarrow \mathbb{S}$ is defined by $q = p + 4$.

6. $\mathbb{R} = \{t\colon t \text{ is a real number}\}$, $\mathbb{S} = \{u\colon u \text{ is a real number}\}$, where $\mathscr{T}\colon \mathbb{R} \leftrightarrow \mathbb{S}$ is defined by $u = t - 7$.

7. $\mathbb{R} = \{h\colon h \text{ is an integer}\}$, $\mathbb{S} = \{k\colon k \text{ is a rational number}\}$, where $\mathscr{T}\colon \mathbb{R} \leftrightarrow \mathbb{S}$ is defined by $k = h/5$.

8. $\mathbb{R} = \{a\colon a \text{ is a real number}\}$, $\mathbb{S} = \{b\colon b \text{ is a real number}\}$, where $\mathscr{T}\colon \mathbb{R} \leftrightarrow \mathbb{S}$ is defined by $b = a/6$.

In Exercises 9–14, determine whether the given conditions yield isome-tries between $\mathbb{R}$ and $\mathbb{S}$.

9. $\mathbb{R} = \{(x, y)\colon y = 3x + 2\}$, $\mathbb{S} = \{(x, y)\colon y = 6x - 4\}$, where $\mathscr{T}\colon \mathbb{R} \leftrightarrow \mathbb{S}$ is defined as follows: Match each point in $\mathbb{R}$ to the point in $\mathbb{S}$ obtained by keeping the same x-coordinate and adding $3x - 6$ to the y-coordinate.

10. $\mathbb{R} = \{(x, y)\colon y = 5x + 7\}$, $\mathbb{S} = \{(x, y)\colon y = 5x - 9\}$, where $\mathscr{T}\colon \mathbb{R} \leftrightarrow \mathbb{S}$ is defined as follows: Match each point in $\mathbb{R}$ to the point in $\mathbb{S}$ obtained by keeping the same x-coordinate and subtracting 16 from the y-coordinate.

11. $\mathbb{R} = \{(x, y)\colon y = 3\}$, $\mathbb{S} = \{(x, y)\colon y = 2\}$, where $\mathscr{T}\colon \mathbb{R} \leftrightarrow \mathbb{S}$ is defined as follows: Match each point in $\mathbb{R}$ to the point in $\mathbb{S}$ obtained by keeping the same x-coordinate and subtracting 1 from the y-coordinate.

12. $\mathbb{R} = \{(x, y)\colon x = -5\}$, $\mathbb{S} = \{(x, y)\colon x = 3\}$, where $\mathscr{T}\colon \mathbb{R} \leftrightarrow \mathbb{S}$ is defined as follows: Match each point $\mathbb{R}$ to the point in $\mathbb{S}$ obtained by keeping the same y-coordinate and adding 8 to the x-coordinate.

13. $\mathbb{R} = \{(x, y): y = x\}, \mathbb{S} = \{(x, y): y = x\}$, where $\mathcal{T} : \mathbb{R} \leftrightarrow \mathbb{S}$ is defined as follows: Match each point in $\mathbb{R}$ to the point in $\mathbb{S}$ obtained by keeping the same x-coordinate and the same y-coordinate.

14. $\mathbb{R} = \{(x, y): y = x\}, \mathbb{S} = \{(x, y): y = x^2\}$, where $\mathcal{T} : \mathbb{R} \leftrightarrow \mathbb{S}$ is defined as follows: Match each point $\mathbb{R}$ to the point in $\mathbb{S}$ obtained by keeping the same x-coordinate and squaring the y-coordinate.

In Exercises 15–20, determine whether the set $\mathbb{S}$ is a translation of the set $\mathbb{R}$.

15. $\mathbb{R} = \{(x, y): y = 3x - 2\}, \mathbb{S} = \{(x, y): y = 3x - 8\}$.

16. $\mathbb{R} = \{(x, y): y = 2x + 5\}, \mathbb{S} = \{(x, y): y = 2x - 7\}$.

17. $\mathbb{R} = \{(x, y): y = 6\}, \mathbb{S} = \{(x, y): y = -4\}$.

18. $\mathbb{R} = \{(x, y): x = -3\}, \mathbb{S} = \{(x, y): x = 2\}$.

19. $\mathbb{R} = \{(x, y): y = x\}, \mathbb{S} = \{(x, y): y = 2x\}$.

20. $\mathbb{R} = \{(x, y): y = x\}, \mathbb{S} = \{(x, y): y = x^2\}$.

In Exercises 21–26, determine whether the set $\mathbb{S}$ is a rotation of the set $\mathbb{R}$.

21. $\mathbb{R} = \{(x, y): y = 5x\}, \mathbb{S} = \{(x, y): y = 4x\}$.

22. $\mathbb{R} = \{(x, y): y = 6x\}, \mathbb{S} = \{(x, y): y = -4x\}$.

23. $\mathbb{R} = \{(x, y): y = 6\}, \mathbb{S} = \{(x, y): y = -4\}$.

24. $\mathbb{R} = \{(x, y): x = -3\}, \mathbb{S} = \{(x, y): y = 2\}$.

25. $\mathbb{R} = \{(x, y): y = x\}, \mathbb{S} = \{(x, y): y = 3\}$.

26. $\mathbb{R} = \{(x, y): y = x\}, \mathbb{S} = \{(x, y): y = x^2\}$.

In Exercises 27–32, determine whether the set $\mathbb{S}$ is a reflection of the set $\mathbb{R}$ with respect to the line $y = x$.

27. $\mathbb{R} = \{(x, y): y = x + 2\}, \mathbb{S} = \{(x, y): y = x - 2\}$.

28. $\mathbb{R} = \{(x, y): x = y - 5\}, \mathbb{S} = \{(x, y): x = y + 5\}$.

29. $\mathbb{R} = \{(x, y): y = 2x + 3\}, \mathbb{S} = \{(x, y): y = 2x - 3\}$.

30. $\mathbb{R} = \{(x, y): x = y\}, \mathbb{S} = \{(x, y): y = x\}$.

31. $\mathbb{R} = \{(x, y): (x + 5)^2 + (y - 7)^2 = 5\}, \mathbb{S} = \{(x, y): (x - 7)^2 + (y + 5)^2 = 5\}$.

32. $\mathbb{R} = \{(x, y): (x + 2)^2 - (y + 6)^2 = -1\}, \mathbb{S} = \{(x, y): (x + 6)^2 - (y + 2)^2 = -1\}$.

In Exercises 33–40, you are given a combination of translations, rotations, and/or reflections. Perform the transformations in the order
given by drawing an initial diagram of the set $\mathbb{R}$, with a drawing
after each transformation. Then change the order of the transformations as indicated and determine whether the order in which these
are performed changes the result.

33. Let $\mathbb{R} = \{(x, y): y = 2x + 3\}$. First: rotate $\mathbb{R}$ 90° counterclockwise
and then rotate 30° clockwise. Second: rotate $\mathbb{R}$ 30° clockwise and
then rotate 90° counterclockwise. Is the final result a rotation of $\mathbb{R}$?

34. Let $\mathbb{R} = \{(x, y): y = 2x + 3\}$. First: translate $\mathbb{R}$ 4 units to the
right and 3 units down and then translate 7 units to the left. Second:
translate 7 units to the left and then translate 4 units to the right and
3 units down. Is the final result a translation of $\mathbb{R}$?

35. Let $\mathbb{R} = \{(x, y): y = x - 2\}$. First: translate $\mathbb{R}$ 3 units to the right
and then rotate 90° clockwise. Second: rotate $\mathbb{R}$ 90° clockwise and
then translate 3 units to the right.

36. Let $\mathbb{R} = \{(x, y): y = x + 3\}$. First: translate $\mathbb{R}$ 5 units to the left
and then rotate 90° counterclockwise. Second: rotate $\mathbb{R}$ 90° counterclockwise and then translate 5 units to the left.

37. Let $\mathbb{R} = \{(x, y): y = -2x\}$. First: reflect $\mathbb{R}$ about the line $y = x$
and then rotate 30° counterclockwise. Second: rotate $\mathbb{R}$ 30° counterclockwise and then reflect about the line $y = x$.

38. Let $\mathbb{R} = \{(x, y): y = 3x + 1\}$. First: reflect $\mathbb{R}$ about the line $y = -x$
and then rotate 60° clockwise. Second: rotate $\mathbb{R}$ 60° clockwise and
then reflect about the line $y = -x$.

39. Let $\mathbb{R} = \{(x, y): x^2 + y^2 = 4\}$. First translate $\mathbb{R}$ 1 unit up, then rotate
45° clockwise, and then reflect about the line $y = -x$. Second: rotate
$\mathbb{R}$ 45° clockwise, then translate 1 unit up, and then reflect about the
line $y = -x$. Third: reflect $\mathbb{R}$ about the line $y = -x$, then rotate
45° clockwise, and then translate 1 unit up. (There are three other
orders possible. List them and make the appropriate drawings.)

40. Let $\mathbb{R} = \{(x, y): x^2 + (y + 2)^2 = 9\}$. First: translate $\mathbb{R}$ 2 units
down, then rotate 45° counterclockwise, and then reflect about the
line $y = x$. Second: reflect $\mathbb{R}$ about the line $y = x$, then rotate 45°
counterclockwise, and then translate 2 units down. Third: rotate $\mathbb{R}$
45° counterclockwise, then translate 2 units down, and then reflect
about the line $y = x$. (There are three other orders possible. List
them and make the appropriate drawings.)

Proofs:

41. Prove that an isometry preserves betweenness.

42. Prove that an isometry preserves the linearity of 3 points.

43. Prove that an isometry on a triangle yields a congruent triangle.

44. A **magnification** is a transformation $\mathcal{T} : \mathbb{R} \leftrightarrow \mathbb{S}$ in which every point in $\mathbb{R}$ with coordinates (x, y) corresponds to a point in $\mathbb{S}$ with coordinates (rx, ry), where $r > 1$. Prove that a magnification preserves similarity of triangles, but not congruence.

45. A **reduction** is a transformation $\mathcal{T} : \mathbb{R} \leftrightarrow \mathbb{S}$ in which every point in $\mathbb{R}$ with coordinates (x, y) corresponds to a point in $\mathbb{S}$ with coordinates (rx, ry), where $0 < r < 1$. Prove that a reduction preserves similarity of triangles, but not congruence.

46. Prove Theorem 8.5. [Hint: Let A and B be points in $\mathbb{R}$ corresponding to their reflections A' and B' in $\mathbb{S}$ with respect to line k. Let P and Q be the midpoints of $\overline{AA'}$ and $\overline{BB'}$, respectively. Show that $\triangle APQ \simeq \triangle A'PQ$ using LL = LL. Explain why $\angle AQP \simeq \angle A'QP$ and $\angle AQB \simeq \angle A'QB'$. Then show that $\triangle AQB \simeq \triangle A'QB'$ using SAS = SAS. It follows that $AB = A'B'$.]

In Exercises 47–50, the circles and squares have their centers at the origin. Use the following definition: An inversion is a transformation $\mathcal{T} : \mathbb{R} \leftrightarrow \mathbb{S}$ in which every point in $\mathbb{R}$ with coordinates (x, y) corresponds to a point in $\mathbb{S}$ with coordinates $(x/(x^2+y^2), y/(x^2+y^2))$, where x and y are not both zero.

47. If $\mathbb{R}$ is a unit circle, find $\mathbb{S}$ and prove that $\mathcal{T}$ is a transformation, and also an isometry.

48. If $\mathbb{R}$ is a square with a unit apothem, find $\mathbb{S}$ and prove that $\mathcal{T}$ is a transformation, but not an isometry.

49. If $\mathbb{R}$ is the interior of a unit circle, excluding the origin, find $\mathbb{S}$ and prove that $\mathcal{T}$ is a transformation, but not an isometry.

50. If $\mathbb{R}$ is the exterior of a unit circle, find $\mathbb{S}$ and prove that $\mathcal{T}$ is a transformation, but not an isometry.

Constructions:

51. Construct the reflection of a triangle across a line k. (Several cases: line k intersects the triangle in 0, 1, or 2 points, or in a line segment.)

52. Construct the reflection of a triangle across a line k. (Several cases: line k intersects the triangle in 0, 1, or 2 points, or in a line segment.)

53. Construct the reflection of a circle P across a line k. (Several cases: line k intersects the circle in 0, 1, or 2 points.)

Chapter 8
Key Terms

Chapter 8
Review Exercises

Fill each blank with the word *always, sometimes,* or *never.*

1. In the ordered pair representing a point, the ordinate _____ precedes the abscissa.

2. A triangle with vertices at $(a, b), (c, b)$, and $(c, -b)$ is _____ a right triangle.

3. The slope of a line is _____ greater than or equal to zero.

4. The coordinates of the midpoint of a line segment can _____ be found by taking one-half of the difference of x-coordinates and one half the difference of y-coordinates.

5. If the slope of a line is m, then the line perpendicular to it _____ has a slope of $-1/m$.

True-False: If the statement is true, mark it so. If it is false, replace the underlined word to make a true statement.

6. The abscissa of a point represents the distance of the point from the <u>x-axis</u>.

7. The slope of the line through $P(4, -3)$ and $Q(4, -5)$ is <u>one</u>.

8. The lines represented by $4x + 3y = 5$ and $8 - 12x - 9y = 0$ are <u>parallel</u>.

9. The points $(-7, 5), (-3, 3), (1, 1)$, and $(5, -1)$ <u>are</u> the vertices of a parallelogram.

10. Given points $A(4, 3)$ and $B(2, -2)$, the distance AB is <u>greater than</u> the distance from the origin to A.

Answer the following questions.

11. Find the distance between $P(5, -8)$ and $Q(12, 8)$.

12. Find the distance of the point $A(-7, -12)$ from the x-axis.

13. Find two points at a distance of 10 units from the origin such that the coordinates of each point are the negatives of each other.

14. Write the equation of the line through the point $P(0, -7)$ and having slope 2.

15. Write the restricted equation of the line segment whose endpoints are $(2, 5)$ and $(-6, -7)$.

16. Write the equation of the vertical line through $P(5, -2)$.

17. Find the slope and y-intercept of the line represented by $3x - 5y = 11$.

18. Write the restricted equations for the triangle whose vertices are $(2, 2)$, $(-5, 3)$, and $(2, -3)$.

19. Show that the quadrilateral with vertices $(-3, 3)$, $(-5, 2)$, $(8, -7)$, and $(10, -2)$ is a parallelogram.

20. Determine whether the parallelogram in Exercise 19 is a rectangle or not.

21. Use coordinate geometry to prove Theorem 5.2: The median to the hypotenuse of any right triangle is half as long as the hypotenuse.

22. If $\mathbb{R} = \{x\colon x$ is a negative integer$\}$, $\mathbb{S} = \{y\colon y$ is a positive integer$\}$, and $\mathcal{T}\colon \mathbb{R} \leftrightarrow \mathbb{S}$ is defined by $y = -x$, then determine whether $\mathcal{T}$ is a transformation.

23. If $\mathbb{R} = \{(x, y)\colon y = 2x - 5\}$, $\mathbb{S} = \{(x, y)\colon y = 2x\}$, and $\mathcal{T}\colon \mathbb{R} \leftrightarrow \mathbb{S}$ is defined as follows: Match each point in $\mathbb{R}$ to the point in $\mathbb{S}$ obtained by keeping the same x-coordinate and adding 5 to the y-coordinate, then determine whether $\mathcal{T}$ is an isometry.

24. If $\mathbb{R} = \{(x, y)\colon y = 3x + 1\}$ and $\mathbb{S} = \{(x, y)\colon y = 2x - 4\}$, determine whether the set $\mathbb{S}$ is a translation of the set $\mathbb{R}$.

25. If $\mathbb{R} = \{(x, y)\colon y = x - 3\}$ and $\mathbb{S} = \{(x, y)\colon y = x + 3\}$, determine whether the set $\mathbb{S}$ is a rotation of the set $\mathbb{R}$.

26. If $\mathbb{R} = \{(x, y)\colon y = -2,$ for $0 \leq x \leq 5\}$ and $\mathbb{S} = \{(x, y)\colon x = 2,$ for $-5 \leq y \leq 0\}$, determine whether the set $\mathbb{S}$ is a reflection of the set $\mathbb{R}$ with respect to the line $y = -x$.

Space Geometry

Plato (427?–347 B.C.)

Courtesy of Historical Pictures Service, Chicago

Historical Note

The Greek philosopher Plato was born about 427 B.C. and died in 347 B.C. He made a connection between Pythagorean geometry and Empedoclean biology, by a mathematical construction of the elements. Starting with the isosceles right triangle and the right triangle in which the hypotenuse is double the shortest side, Plato constructed four of the regular solids: the cube, the tetrahedron, the octahedron, and the icosahedron. These four solids were assumed to be the shapes of the corpuscles of earth, fire, air, and water. These ideas affected the study of physiology and medicine as they were thought to be the basic building blocks of all organic and inorganic compounds.

In Section 1.3, we presented Postulate 1.6, which indicates that no plane contains all points of space. Until now, we have concentrated on geometry in a plane, and have essentially ignored the existence of points outside any given plane. In this chapter, we present topics concerning *three-dimensional* space.

9.1   LINES AND PLANES IN SPACE

In Chapter 3, we defined two distinct lines in a plane as parallel if they do not intersect. Thus, in a plane, two distinct lines must either be parallel or they must intersect. This is not true in space. In fact, we define **skew lines** as two distinct lines that do not intersect and are not parallel. In Figure 9.1, lines k and m are skew lines, where line k is on the top of the *box* and line m is on the front of the box.

However, two distinct *planes* are **parallel** if they do not intersect. There are only two possibilities for planes: they are parallel or they intersect. In Figure 9.1, if the plane passing through the top of the box does not intersect the plane passing through the bottom of the box, then the two planes are parallel.

In order to determine the type of object that we obtain as the intersection of two planes, we need the following postulate.

Figure 9.1

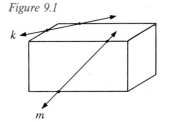

Postulate 9.1	The intersection of two distinct nonparallel planes is a line.

Figure 9.2

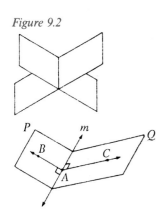

Figure 9.2 illustrates the intersection of two planes, and also the intersection of two half-planes. When two half-planes have a common edge, they form a **dihedral angle**. The edge of intersection of the two half-planes is the **edge** of the dihedral angle. The half-planes are the **sides** of the dihedral angle.

Consider a point A on the edge m of a dihedral angle with half-planes P and Q. Let point B be in half-plane P and point C be in half-plane Q such that ray $AB \perp m$ and ray $AC \perp m$. Then the measure of the dihedral angle is defined to be equal to the measure of $\angle BAC$.

If the union of two distinct half-planes is a plane, then the half-planes are **opposite half-planes**. If two dihedral angles with the same edge are placed such that the sides of one dihedral angle form opposite half-planes with the sides of the other dihedral angle, then the dihedral angles are **vertical dihedral angles**.

Figure 9.3

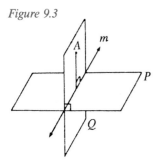

Two planes are **perpendicular** if the measures of their vertical dihedral angles are each 90°. To find the **distance** between a point A and a plane P, illustrated in Figure 9.3, pass a plane Q through point A and perpendicular to plane P and find the distance between the point A and the intersection line m of the two planes. We discussed how to find the distance between a point and a line in Section 1.6.

Similarly, the **distance** between two parallel planes is found by taking any point on one plane and finding the distance between that point and the other plane.

EXERCISE SET 9.1

1. Label the corners of the box in Figure 9.1 A, B, C . . . H. Which line segments forming edges of the box are parallel? Which are intersecting? Which are perpendicular? Which are skew?

In Exercises 2–6, let m be the line of intersection of distinct planes P and Q, and let A be a point on m.

2. How many planes can be drawn through m?

3. If k is a line in Q, with $k \parallel m$, how many planes can be drawn through m and k?

4. How many lines can be drawn through A and perpendicular to m?

5. Can a line be drawn perpendicular to both planes P and Q?

6. How many planes can be drawn through A and perpendicular (a) to m, (b) to P, (c) to both P and Q?

In Exercises 7–11, draw a figure to illustrate the answers to the questions.

7. Must two parallel lines be in the same plane?

8. Must three parallel lines be in the same plane?

9. Under what conditions is a plane T parallel to both of the planes R and S?

10. Point A is not in plane R. How many lines can be drawn through A parallel to R? How many planes?

11. Point A is not in plane R. How many lines can be drawn through A perpendicular to R? How many planes?

In Exercises 12–18, draw figures illustrating each of the following:

12. Vertical dihedral angles

13. A dihedral angle of measure 30°

14. A dihedral angle of measure 120°

15. A dihedral angle of measure 180°

16. Perpendicular planes

17. Opposite half planes

18. The distance from a point A to a plane P, where A is not in P

19. Given a point A on plane P and a point B on plane Q, show that it is possible for the distance between A and B to equal 5, while the distance between P and Q equals 4.

20. In the situation described in Exercise 19, show that it is impossible for the distance between A and B to equal 4 when the distance between P and Q equals 5.

9.2 SOLIDS AND SURFACES

The next definition may seem a little strange, inasmuch as we are defining something called a **convex solid** without defining *solid*. However, the general definition of a solid is unnecessary for this course, as we will discuss only solids that are convex. An example of a non-convex object that most individuals would recognize as being solid is a doughnut.

Figure 9.4

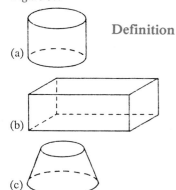

Definition

> A **convex solid** is a set of points S, not all in one plane, such that for any two points A and B in S, all points between A and B are also in S.

Given a solid S, suppose that we have points P and Q in S such that for every point R satisfying $Q-P-R$, R is not in S. Then we say that P is a **surface point** of the solid S. The set of all surface points of S is the **surface** of S. All points of S that are not in the surface of S are in the **interior** of S. Points in space that are not in S are in the **exterior** of S. See Figure 9.4 for several examples of solids and surfaces.

We will now consider a number of different types of solids. Many of these solids should be familiar to the reader.

Figure 9.5

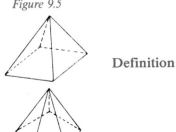

Definition

> A **polyhedron** is a solid whose surface is the union of a finite number of polygonal regions.

Each polygonal region is a **face** of the polyhedron. If the intersection of two faces is a line segment, the line segment is an **edge** of the polyhedron. If the intersection of two edges is a point, the point is a **vertex** of the polyhedron.

Figure 9.6

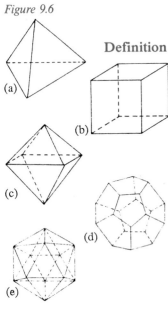

(a)
(b)
(c)
(d)
(e)

Definition

> A **regular polyhedron** is a solid whose faces are all regular congruent polygonal regions.

An example of a polyhedron that need not be regular is a **pyramid**, which is a polyhedron with one face (the **base**), being an arbitrary polygonal region and the other faces being triangular regions that meet in a common point, the **vertex**. See Figure 9.5 for some examples of pyramids. The only possible regular pyramid is described here.

Only five types of *regular* polyhedrons exist. A regular **tetrahedron** (Figure 9.6a), has four equilateral triangular faces. A regular **hexahedron** (Figure 9.6b), also known as a **cube**, has six square faces. A regular **octahedron** (Figure 9.6c), has eight equilateral triangular faces. A regular **dodecahedron** (Figure 9.6d), has twelve regular pentagonal faces. Finally, a regular **icosahedron** (Figure 9.6e), has twenty equilateral triangular faces.

■

Example 1

Find (a) the number of faces of a cube
(b) the number of vertices of a tetrahedron
(c) the number of edges of an octahedron

Solution

The following are determined by examining Figure 9.6.

(a) six faces
(b) four vertices
(c) twelve edges

■

Definition	A **prism** is a polyhedron, two of whose faces are parallel and congruent and whose other faces are parallelograms.

The parallel faces are **bases**. The faces that are not bases are **sides**. A **right prism** has rectangular sides.

Figure 9.7

Definition	A **parallelepiped** is a prism with six faces that are parallelograms.

A rectangular parallelepiped is a fancy name for what the reader would probably call a *box*. A cube is an example of a rectangular parallelepiped. See Figure 9.7 for sketches of prisms.

Definition	A **diagonal of a polyhedron** is a line segment whose endpoints are vertices not in the same face of the polyhedron.

Example 2

How many diagonals does a rectangular parallelepiped have?

Solution

Figure 9.4b is a rectangular parallelepiped. It has four diagonals. For example, one diagonal extends from the front, top, right to the back, bottom, left of the parallelepiped.

There are many solids that are not polyhedrons. Some of these are defined here.

Definition	A **sphere** is the set of all points in space at a given distance from a given point. The distance is the **radius** of the sphere and the point is the **center** of the sphere.

The **interior of a sphere** is the set of all points in space at a smaller distance from the center than the radius. The union of the sphere and its interior forms a solid. This solid is often also called a sphere, which can cause confusion with the previous definition. For our purposes in this text, we use *sphere* when referring to the surface, except when discussing volumes of solids.

Definition

> Let *C* be a circular region and *P* a point not in the plane containing the circular region. **A circular cone** is the union of *C*, *P*, and all points between *P* and each point of *C*. It has **vertex** *P* and **base** *C*.

A circular cone is clearly a solid. The line segment from *P* to the plane containing *C* and perpendicular to the plane containing *C* is the **altitude of the cone**. If the line segment joining *P* to the center of region *C* is perpendicular to the plane containing *C*, then the solid is a **right circular cone**. (See Figure 9.8.) If the line segment joining *P* to the center of the region *C* is *not* perpendicular to the plane containing *C*, then the solid is an **oblique circular cone**. If a region other than a circular region is used, the cone is a **non-circular cone**.

Figure 9.8

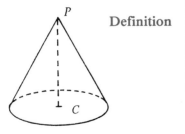

Figure 9.9

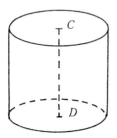

Definition

> Let *C* and *D* be two circular regions in parallel planes. **A circular cylinder** is the solid formed by the union of *C*, *D*, and all points between each point of *C* and *D*. The **bases** of the cylinder are *C* and *D*.

If the segment joining the centers of the circular regions is perpendicular to both planes, then the solid is a **right circular cylinder**. See Figure 9.9. If the segment joining the centers of the circular regions is *not* perpendicular to both planes, the solid is an **oblique circular cylinder**. If a region other than a circular region is used, the cylinder is a **non-circular cylinder**.

In general, an **altitude of a solid** having a base is a line segment that satisfies the following three conditions for the plane containing the base.

(i) It is perpendicular to the plane.
(ii) One endpoint is on the plane.
(iii) The other endpoint is a point, on the solid, farthest from the plane.

EXERCISE SET 9.2

In Exercises 1–6, let *f* be the number of faces, *v* the number of vertices, and *e* the number of edges of the polyhedron.

1. Find *f*, *v*, and *e* for a tetrahedron.

2. Find *f*, *v*, and *e* for a cube.

3. Find *f*, *v*, and *e* for an octahedron.

4. Find *f*, *v*, and *e* for a dodecahedron.

5. Find f, v, and e for an icosahedron.

6. Find $f + v - e$ for each of Exercises 1–5. (According to Euler's Theorem: Every polyhedron, regular or irregular, has the property that $f + v - e = 2$.)

7. How many faces of a tetrahedron meet at each vertex? How many edges meet at each vertex?

8. Answer the questions in Exercise 7 for a cube.

9. Answer the questions in Exercise 7 for an octahedron.

10. Find the number of degrees in the sum of measures of the angles at each vertex of a tetrahedron, a cube, an octahedron, and an icosahedron. (It can be proved that this sum must always be less than $360°$.)

11. How many diagonals does a tetrahedron have?

12. How many diagonals does an octahedron have?

13. How many diagonals does a dodecahedron have?

14. How many diagonals does an icosahedron have?

15. Sketch an irregular tetrahedron.

16. Sketch an irregular hexahedron.

17. Sketch an irregular polyhedron having nine faces.

18. Sketch a pyramid having a hexagon as base. How many faces does it have?

19. Sketch a prism having pentagons as the parallel faces. How many faces does it have? How many diagonals does it have?

20. Sketch a parallelepiped.

21. Sketch a right circular cylinder with altitude three times the radius of the base.

22. Sketch a right circular cylinder with altitude twice the radius of the base.

23. Sketch a circular cylinder that is not a right circular cylinder, with altitude equal to the radius of the base.

24. Sketch a circular cylinder that is not a right circular cylinder, with altitude half the radius of the base.

25. Sketch a right circular cone whose altitude is twice the radius of the base.

26. Sketch a right circular cone whose altitude is half the radius of the base.

9.3 SURFACE AREA

In Chapter 7, we discussed various theorems relating to the areas of polygons. We use these same theorems to find the surface area of polyhedrons.

Definition | The **surface area** of a polyhedron is the sum of the areas of the faces of the polyhedron.

Example 1

Figure 9.10

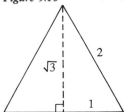

Find the surface area of a regular tetrahedron each of whose edges is 2 inches.

Solution

There are four faces, each of which has the same area. Each face is an equilateral triangle of area $\sqrt{3}$ square inches, by Theorem 7.3. See Figure 9.10. Thus, the surface area is $4\sqrt{3}$ square inches.

Example 2

Figure 9.11

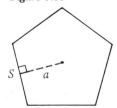

Find the surface area of a regular dodecahedron each of whose faces has apothem a and side s.

Solution

See Figure 9.11. There are twelve faces, each of which has the same area. Each face is a regular pentagon whose perimeter is $5s$. Thus, by Corollary 1 to Theorem 7.3, its area is $5as/2$ square units, and the surface area of the dodecahedron is $30as$ square units.

Although the previous examples demonstrated how we find the surface areas of regular polyhedrons, polyhedrons need not be regular in order for us to find their surface areas.

Example 3

Find the surface area of the rectangular paralellepiped shown in Figure 9.12.

Solution

Figure 9.12

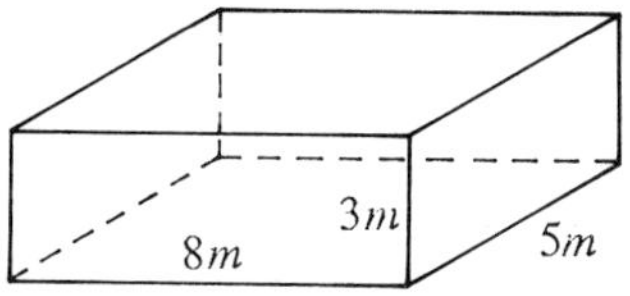

The rectangular parallelepiped has a length of 8 meters, a height of 3 meters, and a width of 5 meters. (The names *length, width,* and *height* are chosen arbitrarily for the edges, although we usually consider a length to be larger than a width.) Using Theorem 7.1, we find that two of the faces have an area of 24 square meters each, two have an area of 40 square meters each, and two have an area of 15 square meters each. Thus, the area of the parallelepiped is 158 square meters.

Consider a circular cone formed by a circular region C and a point P not in the plane containing C. Let D be a polygonal region formed by inscribing a regular polygon in the circle. The union of P, D, and all points between P and the points of D form a pyramid with base D, which is *inscribed* in the cone. Consider what happens as we increase the number of sides of the base, as in Figure 7.9. The perimeter of the polygon approaches the circumference of the circle as a limiting value, and the surface area of the pyramid approaches a number we define to be the *surface area of a cone*. The *surface area of a circular cylinder* is defined similarly, using prisms instead of pyramids.

The **lateral surface area** of a cone is the cone's surface area minus the area of its base. The *lateral surface area* of a circular cylinder is the surface area of the circular cylinder minus the area of its two bases.

In defining the *surface area* of a sphere, we inscribe polyhedrons in the sphere, but we must be very careful. The largest regular polyhedron has twenty faces. Therefore, we cannot continue to increase the number of faces of regular polyhedrons inscribed in a sphere. We must use irregular polyhedrons. In doing so, we must be careful that these polyhedrons become *closer* to the sphere as we increase the number of faces. The limiting value of the surface area of the polyhedrons is the surface area of the sphere.

We state the following theorems without proof.

Theorem 9.1

$(S = \pi rs)$ The lateral surface area of a right circular cone is the product of π, the radius of the circle, and the distance from the vertex to any point on the circle.

The distance from the vertex of a right circular cone to any point on the circle, referred to in Theorem 9.1, is the **slant height** of the cone.

Example 4

Figure 9.13

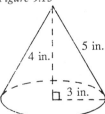

Find the surface area of a right circular cone with radius 3 inches and altitude 4 inches.

Solution

See Figure 9.13. Using the Pythagorean Theorem, we find that the distance from the vertex to the circle is 5 inches. Thus, the lateral surface area is 15π square inches and the area of the base is 9π square inches. It follows that the total surface area is 24π square inches.

Theorem 9.2	**(S = 2πra)** The lateral surface area of a right circular cylinder is two times the product of π, the radius of the circle, and the length of the altitude of the cylinder.

Example 5

Find the surface area of a right circular cylinder with radius 2 feet and altitude 7 feet.

Solution

The lateral surface area is 28π square feet and the area of each base is 4π square feet. Thus, the total surface area is 36π square feet.

Theorem 9.3	**(S = 4πr²)** The surface area of a sphere is four times the product of π and the square of the radius.

■

Example 6

Find the surface area of a sphere of radius 5 miles.

Solution

By Theorem 9.3, its surface area equals 100π square miles.

■

EXERCISE SET 9.3

1. Find the surface area of a regular tetrahedron if each of its edges is 3 meters. Compare this area with that of a regular octahedron, each of whose edges is 3 meters.

2. Find the surface area of a cube if each edge is 16/3 inches.

3. Find the surface area of a rectangular parallelepiped having length 12 inches, width 10 inches, and height 5 inches.

4. Find the surface area of an icosahedron if each edge is 8 centimeters.

5. Find the lateral surface area of a cone with radius of the base 5 centimeters and altitude 8 centimeters.

6. Find the entire surface area of a cone if the radius of the base is 6 inches and the altitude is 8 inches.

7. Find the lateral surface area of a right circular cylinder with radius of base 7 inches and altitude 10 inches.

8. Find the entire surface area of a cylindrical tin can if the radius of the bottom is 8 centimeters and the height is 15 centimeters.

9. The radius of the earth is approximately 4000 miles. If the earth were an exact sphere (totally smooth), what would be the approximate surface area? Look up an approximation of the radius of the earth to the nearest mile and find the surface area. How do the two approximations compare?

10. Consider an ice cream cone with a hemisphere (half of a sphere) of ice cream on top. If the diameter of the top of the cone is 3 inches and the height of the cone is 5 inches, find the entire surface area of the cone and ice cream.

11. Sketch the figure and find the entire surface area of a right prism whose parallel faces are equilateral triangles with sides of 6 inches, and whose lateral faces are squares.

12. The original dimensions of the Great Pyramid of Cheops in Egypt were approximately 750 feet along each side of the square base and 481 feet in height. If the sides had been triangles, what would the surface area have been?

9.4 VOLUME

In this section, we discuss postulates and theorems related to volumes of solids. Because of the elementary nature of this text, we will not prove the theorems, but we call them theorems so the reader will be aware that proofs are possible.

Given a cube whose edges are each 1 unit long, we say the **volume** of the cube is 1 **cubic unit**. For example, a cube whose edges are each 1 mile long has a volume of 1 cubic mile. The volume of a cube whose edges are each 1 centimeter long is 1 cubic centimeter.

The following postulate extends the concept of volume to a larger set of solids.

Postulate 9.2	The volume of a rectangular parallelepiped equals the area of a base times the distance to its parallel base.

Example 1

Find the volume of the rectangular parallelepiped shown in Figure 9.14.

Solution

Figure 9.14

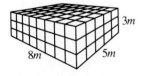

The rectangular parallelepiped has a volume of 120 cubic meters. Any two of the parallel faces of the parallelepiped may be considered to be the bases. For example, if we consider the base to be 8 meters by 5 meters, then we compute the volume by multiplying 40 square meters by 3 meters. If instead, we consider the base to be 3 meters by 5 meters, then we compute the volume by multiplying 15 square meters by 8 meters.

In order to compute the volumes of several of the solids we have already discussed, we need a postulate called **Cavalieri's Principle.** As groundwork for this postulate, we give the following definition.

Definition

> A **cross section** is the nonempty intersection of a solid and a plane. If a plane intersects more than one solid, the cross sections are **corresponding cross sections.**

Postulate 9.3

> Two solids have the same volume if every plane parallel to a given plane cuts the two solids such that corresponding cross sections have equal areas.

This principle can be illustrated by placing two identical decks of playing cards on a flat surface. Neatly stack the cards of one deck on top of each other, while stacking the cards of the other deck on top of each other, but with the edges of different cards pointing in different directions. Ignoring the thickness of the cards for illustrative purposes, each card represents a cross section of the solid in which it lies. Thus, although the two solids have different shapes, they have the same volume. The reader should be able to demonstrate this principle using coins, such as quarters, nickels, and dimes, to show that it is not necessary for the cross sections in one solid to have the same area as the other cross sections in the same solid.

Theorem 9.4

> **(V = Ba)** The volume of a prism or a circular cylinder equals the area of a base times the length of the altitude.

Example 2

Find the volume of a prism whose base is a regular hexagon each of whose edges is 8 inches and whose altitude is 10 inches.

Solution

By Corollary 1 to Theorem 7.3, the area of the base is $96\sqrt{3}$ square inches, so the volume of the prism is $960\sqrt{3}$ cubic inches.

■

Example 3

Find the volume of a circular cylinder whose base is a circle of radius 8 inches and whose altitude is 10 inches.

Solution

The area of the base is 64π square inches and the volume of the circular cylinder is 640π cubic inches.

■

Theorem 9.5	**(V = Ba/3)** The volume of a pyramid or a circular cone equals one third the area of the base times the length of the altitude.

Example 4

Find the volume of a pyramid if its altitude is 5 inches and its base is a square whose edges are 9 inches.

Solution

The area of the base is 81 square inches, so the volume of the pyramid is 27 square inches times 5 inches, with a result of 135 cubic inches.

■

Example 5

Find the volume of a cone whose base is a circle of radius 9 inches and whose altitude is 5 inches.

Solution

The area of the base is 81π square inches, so the volume of the circular cone is 135π cubic inches.

■

Theorem 9.6	$(V = 4\pi r^3/3)$ The volume of a sphere equals four thirds the product of π and the cube of the radius.

■

Example 6

Find the volume of a sphere of radius 6 meters.

Solution

By Theorem 9.6, its volume is $4\pi/3$ times 216 cubic inches, with a result of 288π cubic inches.

■

EXERCISE SET 9.4

1. Find the volume of a rectangular parallelepiped having dimensions 20 meters, 7 meters, and 3 meters.

2. If the three dimensions of a parallelepiped are each doubled, find the ratio of the new volume to the old.

3. Answer Exercise 2, if each dimension is tripled.

4. The base of a prism is a parallelogram whose sides are 4 centimeters and 10 centimeters, and form an angle of measure 60°. If the altitude of the prism is 5 centimeters, what is its volume?

5. Find the volume of a right circular cylinder, if the radius of the base is 12 meters and the altitude is 8 meters.

6. Find the volume of a pyramid having a square base 7 inches on each side and having altitude 12 inches.

7. A pyramid has a hexagon for its base, with each side 4 centimeters. If its altitude is 5 centimeters, find the volume.

8. Find the volume of a circular cone, if the radius of the base is 5 feet, and the altitude is 12 feet.

9. Write a formula for the volume of a right circular cone in terms of radius r, if the radius and altitude are equal.

10. Find the volume of a sphere, if the diameter is 10 centimeters.

11. Find the approximate volume of the earth by assuming that the earth is a sphere of radius 4000 miles. Look up an approximation of the radius of the earth to the nearest mile and find the volume. How do the two approximations compare?

12. What is the ratio of the volume of two spheres, if the radius of one is twice the radius of the other?

13. Find the volume of the ice cream cone and ice cream in Exercise Set 9.3, No. 10.

14. A rocket has the shape of a right circlular cylinder topped by a right circular cone of the same radius. Find the volume of the rocket, if the altitude of the cylinder is 250 feet, the altitude of the cone is 140 feet, and the radius is 60 feet.

15. A tank has the shape of a right circular cylinder whose length is 15 feet and whose diameter is 4 feet, with a hemisphere at each end of the cylinder. Find the volume of the tank.

Chapter 9
Key Terms

Chapter 9
Review Exercises

Fill each blank with the word *always,* *sometimes,* **or** *never.*

1. Two nonintersecting lines in space are _____ parallel.

2. Two nonintersecting planes are _____ parallel.

3. Given points A and B in a convex solid, a point C between A and B is _____ in the exterior of the solid.

4. A surface point of a solid is _____ a point in the solid.

5. A prism is _____ a rectangular parallelepiped.

6. The lateral surface area of a circular cone is _____ equal to its total area.

7. The lateral surface area of a circular cylinder _____ equals the sum of the areas of its bases.

8. Two solids _____ have the same volume, if every plane parallel to a given plane cuts the two solids so that corresponding cross sections have unequal areas.

9. Given two solids with the same base, one of which is a right circular cylinder and the other a right circular cone, they _____ have the same volume.

10. A regular tetrahedron _____ has hexagonal faces.

True-False: If the statement is true, mark it so. If it is false, replace the underlined word to make a true statement.

11. Two distinct planes that do not intersect are <u>parallel</u>.

12. When two <u>planes</u> intersect along their edges, they form a dihedral angle.

13. If two lines in space do not intersect, they are either parallel or <u>perpendicular</u>.

14. There are exactly <u>five</u> types of regular polyhedrons.

15. A regular dodecahedron has twelve regular <u>hexagonal</u> faces.

16. The distance between two <u>intersecting</u> planes is found by taking any point on one plane and finding the distance between that point and the other plane.

17. A plane <u>can</u> be perpendicular to both of two intersecting planes.

18. A regular hexahedron has <u>eight</u> faces.

19. The two parallel faces of a prism <u>may be</u> congruent.

20. A dodecahedron has <u>more</u> faces than an icosahedron.

Answer the following questions.

21. In space, through a point outside a given line, how many lines are there that are parallel to the given line?

22. In space, through a point outside a given line, how many lines are there that are skew to the given line?

23. If each edge of a cube is 3 inches, how many of these small cubes would be needed to form a cube having an edge of 12 inches?

24. How many faces of a regular icosahedron meet at each vertex?

25. Find the surface area of a regular tetrahedron, each of whose edges is 5 centimeters long.

26. Find the surface area of a regular icosahedron, each of whose edges is 20 miles long.

27. Which solid has the greater volume, a right circular cone with altitude 10 inches and radius of the base 6 inches or a pyramid having an altitude of 10 inches and a square base with a side of 10 inches?

28. Find the lateral surface area of a right circular cone whose base has radius 7 inches, if the distance from the vertex of the cone to its base is 10 inches.

29. Find the total surface area of a right circular cylinder whose bases have radii 9 feet and whose altitude is 20 feet.

30. Find the surface area of a sphere whose radius is 3900 miles.

31. If a rectangular parallelepiped has a volume of 127 cubic meters, a length of 3 meters, and a width of 9 meters, find its height.

32. Suppose that a solid has a volume of 192.3 cubic centimeters and that every plane parallel to a given plane cuts this solid and a second solid so that corresponding cross sections have equal areas. What is the volume of the second solid?

33. Suppose that a prism has a base area of 29 square yards and an altitude of length 15 yards. Find the volume of the prism.

34. Suppose that a pyramid with altitude 13 feet has a base with 17 sides whose area is 23 square feet. Find the volume of the pyramid.

35. Find the volume of a sphere whose radius is 93 million miles.

Non-Euclidean Geometries

Karl Friedrich Gauss (1777–1855)

Historical Note

Karl Friedrich Gauss (1777–1855) was known as the "Prince of Mathematicians." At the age of three, he was able to correct his father's errors in payroll computations. At the age of twelve, he questioned the foundations of Euclidean geometry. By the age of sixteen, Gauss had developed some of the first ideas of a non-Euclidean geometry; however, he was not interested in publishing his results. At the age of nineteen, he proved that the construction of a seventeen-sided regular polygon was possible. This achievement convinced him to become a mathematician rather than a philosopher. Accuracy was extremely important in Gauss's mathematical analyses, and he questioned long-standing demonstrations in geometry. Gauss knew of the works of Lobachevsky and Bolyai in what came to be known as non-Euclidean geometry. He developed his results independently, however, and was probably the first person to use the term *non-Euclidean.*

In Chapter 3, we indicated that Postulate 3.1, the Parallel Postulate, caused a great deal of controversy for many centuries. Mathematicians who attempted to prove it, and thereby make it a theorem, failed to do so. Subsequently, mathematicians have decided that the statement must be independent of the other postulates and, therefore, cannot be proved. They have demonstrated this fact by developing models of other geometries (the non-Euclidean geometries) that satisfy all the postulates with the exception of the Parallel Postulate.

10.1 HYPERBOLIC GEOMETRY

In this section, we discuss a model of **hyberbolic geometry**; in the next section, we present a model of **elliptic geometry**.

Two mathematicians, Nikolai Ivanovich Lobachevsky (1792–1856), in Russia, and Johann Bolyai (1802–1860), in Hungary, developed the following postulate.

L-B Postulate

> Through a point outside a given line there are infinitely many parallels to the line.

The following is a fairly simple model of a **non-Euclidean geometry** that satisfies the L-B postulate, given by the Frenchman, Henri Poincaré (1854–1912).

Remember that the words *plane* and *line* are undefined terms. We have been representing planes and lines by figures as in Figure 1.1. Poincaré represented a plane by using the interior of a circle. Points on or outside the circle are ignored.

Lines were represented by Poincaré in a manner that requires a definition of orthogonal circles. Two circles are **orthogonal** if their tangents at the points of intersection are perpendicular. Thus, in Figure 10.1, circle *P* and circle *Q* are orthogonal if the tangents to circle *P* and circle *Q* at *R* are perpendicular and the tangents to circle *P* and circle *Q* at *S* are perpendicular.

We will denote Poincaré's *plane* by *the interior of circle P*. Lines are represented by Poincaré as either the intersection of the interior of circle *P* with a diameter of circle *P* or by the intersection of the interior of circle *P* with circles orthogonal to circle *P*. In this representation of lines, the lines do not have endpoints, just as in Euclidean geometry. In Figure 10.2, see line *AB*, line *CD*, and line *EF*.

Figure 10.1

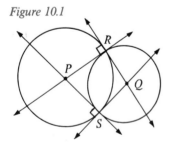

Figure 10.2

Figure 10.3

It is not difficult to show that the system developed under this model satisfies all the postulates of **Euclidean geometry**, other than the Parallel Postulate. Since this textbook is an introductory one, we limit ourselves to an intuitive discussion of the Poincaré model.

Since two lines are parallel if they do not intersect, it is easy to see that in this system, there are infinitely many parallel lines through a point not on a line. In Figure 10.2, line *CD* and line *EF* both pass through point *Q* and both are parallel to line *AB*. In Figure 10.3, we see line *AB* and four lines through point *P* that do not intersect line *AB*.

Figure 10.4

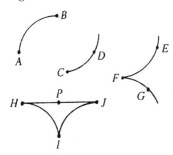

In this geometric model, objects such as line segments, rays, angles, and polygons are still defined in the same way as before, but they may have a different appearance. In Figure 10.4, we see line segment *AB*, ray *CD*, angle *EFG* and triangle *HIJ*. We do not draw arrows on the ends of the rays; instead, we omit endpoints.

Although angles are defined as in Euclidean geometry, and the Degree Protractor Postulate can still be applied, we need to know how to determine the measure of an angle in this new geometry. Given angle *ABC* in Figure 10.5, we form angle *DBE* in the old geometry by using the two tangent rays at the vertex *B*. We then define m∠*ABC* to be equal to m∠*DBE* of the old geometry.

Figure 10.5

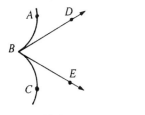

Using this interpretation of measure, the sum of the measures of the angles of a triangle is always *less than* 180°. In Figure 10.6, the objects drawn with solid-line segments represent triangles in our new geometry. The broken-line segments indicate the triangles in Euclidean geometry that have the same vertices. Notice that the measures of the angles of the non-Euclidean triangles are less than the measures of the angles of the Euclidean triangles.

Many other interesting differences exist between Euclidean geometry and hyperbolic geometry. For example, rectangles do not exist in hyperbolic geometry.

Figure 10.6

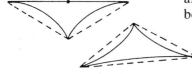

EXERCISE SET 10.1

Answer the following questions for the Poincaré model of hyperbolic geometry.

1. Through a point outside a line, how many parallels are there to the line?

2. Each pair of lines intersects in how many points?

3. How many right angles may a triangle have?

4. If two different lines are parallel to a third line, are they parallel to each other?

5. Through a point outside a line, how many perpendiculars are there to the line?

6. Which polygon has the least number of sides?

7. What can be said about the sum of the measures of the angles of a triangle?

8. Is it true that for three points on a line, exactly one is between the other two?

9. Describe the relationship between the measure of an exterior angle of a triangle and the sum of the measures of the opposite interior angles.

10. Draw a diagram of a line, a line segment, a ray, an angle, and a triangle.

10.2 ELLIPTIC GEOMETRY

The mathematician Georg Friedrich Bernhard Riemann (1826–1866), in Germany, developed the following postulate.

R Postulate

> Through a point outside a given line there are no parallels to the line.

We now consider a **spherical model** for Riemannian geometry. The *plane* in the old geometry will be a sphere in the new geometry. This is not an unrealistic model when we consider that the world on which we live is approximately spherical.

To describe a *line* in the new geometry, we need the concept of a great circle on a sphere. Consider the intersection of a sphere and a plane (in the old geometry), where the plane passes through the center of the sphere. This intersection is a circle, called a **great circle** of the sphere. Great circles determine the shortest path between points on a sphere. They are, therefore, used to determine airplane flight paths. Given a point A and another point B on a sphere, the shortest path between the points A and B, where the path is on the sphere, can be found as follows.

Figure 10.7

Consider the plane passing through the points *A* and *B* and the center of the sphere. Remember that by Postulate 1.4, any three points not on a line determine a unique plane. Intersecting this plane with the sphere will result in a great circle containing the points *A* and *B*. These points divide the circle into two arcs. The distance between *A* and *B* on the sphere will be the length of the shorter arc. On a spherical model of the earth, for example, the shortest route between Albany, New York, and Boulder, Colorado, can be determined by a great circle passing through the two cities. Thus, in practice, a great circle on the earth has the fundamental property that we expect to find in a *line*—namely, that a line determines the shortest distance between two points. See Figure 10.7.

Figure 10.8

Upon careful inspection of great circles on a sphere, we find that any two great circles intersect. (The reader should look not at arbitrary circles on the sphere, but at those that lie on planes passing through the center of the sphere.) Thus, no parallel lines exist in Riemann's geometry. See Figure 10.8.

Figure 10.9

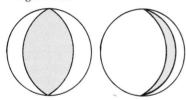

The geometric model in Figure 10.8 satisfies all the postulates of Euclidean geometry with the exception of the Parallel Postulate. There are, however, some interesting differences in our concepts of what some of the geometrical figures look like. For example, lines are *closed*—that is, if you walk far enough along a great circle, you will reach the point from which you started. Furthermore, in this geometry, a ray and a line are identical, and two-sided closed figures exist (two-gons). See Figure 10.9.

Figure 10.10

The measure of angles in elliptic geometry is found using a procedure similar to that of hyperbolic geometry. We find that the sum of the measures of the angles of a triangle is always *greater than* 180°. In particular, there are triangles with three right angles! See Figure 10.10.

EXERCISE SET 10.2

Answer the same questions as in Exercise Set 10.1, for the spherical model of elliptic geometry.

Chapter 10
Key Terms

Chapter 10
Review Exercises

In Exercises 1-10; let P = the Poincaré model of hyperbolic geometry; let S = the spherical model of elliptic geometry; and let E = Euclidean geometry. Indicate in which of the geometries the following statements are true. A statement may be true in more than one geometry. If a statement is not true in any of the above geometries, give the answer as N (N = none of these).

1. The measure of an exterior angle of a triangle may be less than the sum of the measures of the opposite interior angles.

2. There are no parallel lines.

3. Each pair of lines intersects in at most one point.

4. If three points are on a line, exactly one point is between the other two.

5. A triangle may have three right angles.

6. The sum of the measures of the angles of a triangle may range from 180° to 720°.

7. Two different lines may be parallel to a third line, in which case they are parallel to each other.

8. There are no line segments.

9. There exist two-sided polygons.

10. Through a point outside a line there is exactly one perpendicular to the line.

11. Discuss the degree measure of the sum of the angles of a triangle for hyperbolic and elliptic geometries.

12. Do two points determine a line in elliptic geometry? Explain.

13. Draw a quadrilateral in hyperbolic geometry.

14. Draw a triangle which has three right angles (in elliptic geometry).

15. Consider a line in hyperbolic geometry. By the Ruler Postulate, there is a one-to-one correspondence between the set of all real numbers and the set of all points on a line. Assuming that this postulate holds, give an example of such a correspondence and thereby show that lines in hyperbolic geometry are infinitely long.

Appendix A
Algebraic Axioms

REAL NUMBER AXIOMS

Let a, b, and c be real numbers.

1. **Closure**
   If $a \in R$ and $b \in R$, then $a + b \in R$ and $ab \in R$.

2. **Associativity**
   If $a \in R$, $b \in R$, and $c \in R$, then $a + b + c = (a + b) + c = a + (b + c)$ and $abc = (ab)c = a(bc)$.

3. **Commutativity**
   If $a \in R$ and $b \in R$, then $a + b = b + a$ and $ab = ba$.

4. **Distributivity**
   If $a \in R$, $b \in R$, and $c \in R$, then $a(b + c) = ab + ac$.

5. **Additive Identity**
   The set R contains a unique element 0 such that for all elements a in R, $0 + a = a + 0 = a$.

6. **Multiplicative Identity**
   The set R contains a unique element 1 such that for all elements a in R, $1 \cdot a = a \cdot 1 = a$.

7. **Additive Inverse**
   For each element a in R, there exists a unique element in R denoted by $-a$ such that $a + (-a) = (-a) + a = 0$.

8. **Multiplicative Inverse**
   For each element $a \neq 0$ in R, there exists a unique element in R denoted by $1/a$ such that $a \cdot (1/a) = (1/a) \cdot a = 1$.

AXIOMS OF EQUALITY

Let a, b, and c be real numbers.

1. **Reflexivity**
   $a = a$.

2. **Symmetry**
   If $a = b$, then $b = a$.

3. **Transitivity**
   If $a = b$ and $b = c$, then $a = c$.

4. **Substitution**
   If $a = b$, then in any expression containing b, we may replace b by a.

5. **Addition**
   If $a = b$, then $a + c = b + c$.

6. **Multiplication**
   If $a = b$, then $ac = bc$.

AXIOMS OF INEQUALITY

Let a, b, and c be real numbers. We write $b < a$ whenever $a - b$ is positive. We may write $a > b$ instead of $b < a$.

1. **Trichotomy**
   Exactly one of the following statements is true: $a < b$, $a = b$, $a > b$.

2. **Transitivity**
   If $a < b$ and $b < c$, then $a < c$.

3. **Addition**
   If $a < b$, then $a + c < b + c$.

4. **Multiplication**
   If $c > 0$ and $a < b$, then $ac < bc$.
   If $c < 0$ and $a < b$, then $ac > bc$.

Appendix B
List of Postulates

Postulate 1.1   Every line contains at least two distinct points.

Postulate 1.2   Any two distinct points in space have exactly one line that contains them.

Postulate 1.3   Every plane contains at least three distinct points, not all on one line.

Postulate 1.4   Any three distinct points in space not on one line have exactly one plane that contains them.

Postulate 1.5   For any two distinct points in a plane, the line containing the points is also in the plane.

Postulate 1.6   No plane contains all points of space.

Postulate 1.7   (Ruler Postulate)   There is a one-to-one correspondence between the set of all real numbers and the set of all points on a line.

Postulate 1.8   (Degree Protractor Postulate)   Let $\overrightarrow{AB}$ be a ray on the edge of a half-plane H. For every ray $\overrightarrow{AP}$ such that $P \in H$ and $P \notin \overrightarrow{AB}$, there is exactly one real number r such that $0 < r \le 180$.

Postulate 1.9   (Angle Addition Postulate)   If $D \in \text{int}\angle BAC$, then $m\angle BAD + m\angle DAC = m\angle BAC$.

Postulate 2.1   (SAS = SAS)   If two sides and the included angle of one triangle are congruent to the corresponding parts of another triangle, then the triangles are congruent.

Postulate 2.2   (ASA = ASA)   If two angles and the included side of one triangle are congruent to the corresponding parts of another triangle, then the triangles are congruent.

Postulate 2.3   (SSS = SSS) If three sides of one triangle are congruent to the corresponding parts of another triangle, then the triangles are congruent.

Postulate 2.4   A statement and its contrapositive are either both true or both false.

Postulate 3.1   (Parallel Postulate) Through a point outside a given line there is exactly one parallel to the line.

Postulate 4.1   (AAA ~ AAA) If the three angles of one triangle are congruent to the three angles of another triangle, then the triangles are similar.

Postulate 7.1   (Area Postulate) To every region there corresponds exactly one positive real number.

Postulate 7.2   If two triangles are congruent, then their areas are equal.

Postulate 7.3   (Area Addition Postulate) If the intersection of the interiors of two regions is empty, then the area of the union of the regions equals the sum of the areas of the regions.

Postulate 7.4   ($A = s^2$) The area of a square equals the square of the length of a side.

Postulate 7.5   The ratio of a circle's circumference to its diameter is the same for any circle.

Postulate 9.1   The intersection of two distinct nonparallel planes is a line.

Postulate 9.2   The volume of a rectangular parallelepiped equals the area of a base times the distance to its parallel base.

Postulate 9.3   Two solids have the same volume if every plane parallel to a given plane cuts the two solids such that corresponding cross sections have equal areas.

L-B Postulate   Through a point outside a given line there are infinitely many parallels to the line.

R Postulate   Through a point outside a given line there are no parallels to the line.

Appendix C
List of Theorems

Theorem 2.1 Supplements of congruent angles are congruent.

Theorem 2.2 Complements of congruent angles are congruent.

Theorem 2.3 Vertical angles are congruent.

Theorem 2.4 If two sides of a triangle are congruent, the angles opposite these sides are congruent.

Theorem 2.5 If two angles of a triangle are congruent, the sides opposite the angles are congruent.

Theorem 2.6 An exterior angle of a triangle is greater in measure than either remote interior angle.

Theorem 2.7 There is exactly one perpendicular to a line that contains a point not on the line.

Theorem 3.1 If two angles of a triangle are not congruent, the sides opposite these angles are not congruent.

Theorem 3.2 If two sides of a triangle are not congruent, the angles opposite these sides are not congruent, and the angle with larger measure is opposite the longer side.

Theorem 3.3 If two angles of a triangle are not congruent, the sides opposite these angles are not congruent, and the longer side is opposite the angle with larger measure.

Theorem 3.4 Two lines in a plane perpendicular to a third line in the same plane are parallel.

Theorem 3.5 If a line in a plane is perpendicular to one of two parallel lines in the same plane, then it is also perpendicular to the other.

301

Theorem 3.6   If two lines are cut by a transversal so that alternate interior angles are congruent, then the lines are parallel.

Theorem 3.7   If two parallel lines are cut by a transversal, then the alternate interior angles are congruent.

Theorem 3.8   Corresponding angles formed by two lines and a transversal are congruent if and only if the lines are parallel.

Theorem 3.9   Alternate exterior angles formed by two lines and a transversal are congruent if and only if the lines are parallel.

Theorem 3.10   Interior angles formed by two lines and a transversal, such that the angles are on the same side of the transversal, are supplementary if and only if the lines are parallel.

Theorem 3.11   The sum of the measures of the angles of a triangle is $180°$.

Theorem 3.12   An exterior angle of a triangle is equal in measure to the sum of the remote interior angles.

Theorem 3.13   The sum of the measures of the interior angles of an n-gon is $(n-2)180°$.

Theorem 3.14   The sum of the measures of the exterior angles of a polygon, one at each vertex, is $360°$.

Theorem 3.15   A diagonal divides a parallelogram into two congruent triangles.

Theorem 3.16   If the opposite angles or the opposite sides of a quadrilateral are congruent, the quadrilateral is a parallelogram.

Theorem 3.17   Two parallel lines are everywhere equidistant.

Theorem 3.18   The diagonals of a quadrilateral bisect each other if and only if the quadrilateral is a parallelogram.

Theorem 3.19   Two consecutive angles of a parallelogram are supplementary.

Theorem 3.20   Two distinct lines parallel to a third line are parallel to each other.

Theorem 3.21   If three or more parallel lines intercept congruent segments on one transversal, they intercept congruent segments on every transversal.

Theorem 3.22   If a segment joins the midpoints of two sides of a triangle, then it is parallel to the third side and has half the length of the third side.

Theorem 3.23   Base angles of an isosceles trapezoid are congruent.

Theorem 3.24   The median of a trapezoid is parallel to both bases and its length is equal to half the sum of the lengths of the bases.

Theorem 4.1   (AA ~ AA) If two angles of one triangle are congruent to two angles of another triangle, then the triangles are similar.

Theorem 4.2   A line parallel to one side of a triangle that intersects the other two sides in distinct points divides these two sides into proportional segments.

Theorem 4.3   A line that intersects two sides of a triangle in distinct points and divides these two sides into proportional segments is parallel to the third side.

Theorem 4.4   (SAS ~ SAS) If two pairs of corresponding sides of two triangles are proportional and the included angles are congruent, then the two triangles are similar.

Theorem 4.5   (SSS ~ SSS) If corresponding sides of two triangles are proportional, then the two triangles are similar.

Theorem 5.1   (HL = HL) If the hypotenuse and a leg of one right triangle are congruent to the corresponding parts of another right triangle, then the two right triangles are congruent.

Theorem 5.2   The median to the hypotenuse of any right triangle is half as long as the hypotenuse.

Theorem 5.3   In any right triangle, the altitude to the hypotenuse forms two right triangles that are similar to each other and to the original triangle.

Theorem 5.4   Given a right triangle, the altitude to the hypotenuse divides the hypotenuse into two segments such that (1) the altitude is the geometric mean of these segments, (2) each leg is the geometric mean of the hypotenuse and the segment of the hypotenuse adjacent to the leg.

Theorem 5.5   (Pythagorean Theorem) The square of the length of the hypotenuse of a right triangle equals the sum of the squares of the lengths of the legs.

Theorem 5.6   In a right triangle, if an acute angle has a measure of 30°, then the leg opposite this angle is half as long as the hypotenuse.

Theorem 5.7   If the sum of the squares of the lengths of two sides of a triangle equals the square of the length of the third side, then the triangle is a right triangle, with the right angle opposite the third side.

Theorem 6.1   A line perpendicular to a radius at a point on a circle is a tangent of the circle.

Theorem 6.2   A tangent of a circle is perpendicular to the radius of the circle with an endpoint on the point of tangency.

Theorem 6.3   If a secant containing the center of a circle is perpendicular to a chord, then it bisects the chord.

Theorem 6.4   If a secant containing the center of a circle bisects a chord that is not a diameter, then the secant is perpendicular to the chord.

Theorem 6.5   In the plane of a circle, the perpendicular bisector of a chord contains the center of the circle.

Theorem 6.6   In the same circle or in congruent circles, chords equidistant from the center are congruent.

Theorem 6.7   In the same circle or in congruent circles, any two congruent chords are equidistant from the center.

Theorem 6.8   Given two chords in the same circle or in congruent circles, the chord nearer the center is longer.

Theorem 6.9   (Arc Addition Theorem)   If C is a point of $\overset{\frown}{AB}$, then $m\overset{\frown}{AB} = m\overset{\frown}{AC} + m\overset{\frown}{CB}$.

Theorem 6.10   The measure of an inscribed angle is half the measure of its intercepted arc.

Theorem 6.11   The measure of an angle whose vertex is on a circle, such that one ray is a tangent and the other a secant, equals half the measure of its intercepted arc.

Theorem 6.12   Given a pair of vertical angles with the vertex in the interior of a circle, the measure of each angle equals half the sum of the measures of the two intercepted arcs.

Theorem 6.13   Given an angle with the vertex in the exterior of a circle, such that each of the rays of the angle intersect the circle, the measure of the angle equals half the absolute value of the difference of the measures of the intercepted arcs.

Theorem 6.14   In the same circle or in congruent circles, two chords are congruent if and only if the corresponding arcs are congruent.

Theorem 6.15   The two tangent segments from a point to a circle are congruent.

Theorem 6.16   If two chords of a circle intersect in the interior of the circle, the product of the lengths of the segments on one chord equals the product of the lengths of the segments on the other chord.

Theorem 6.17   If $\{B, C, D, E\} \subset \odot P$ and point A satisfies $A-B-C$ and $A-D-E$, then $AB \cdot AC = AD \cdot AE$.

Theorem 6.18   If $\{B, C, D\} \subset \odot P$ and $\overleftrightarrow{AB}$ is a tangent of $\odot P$ where point A satisfies $A-C-D$, then $(AB)^2 = AC \cdot AD$.

Theorem 7.1 $(A = sa)$ The area of a rectangle is the product of the length of any side and the length of the altitude to that side.

Theorem 7.2 $(A = sa)$ The area of a parallelogram is the product of the length of any side and the length of the altitude to that side.

Theorem 7.3 $(A = sa/2)$ The area of a triangle is half the product of the length of any side and the length of the altitude to that side.

Theorem 7.4 $[A = a(b_1 + b_2)/2]$ The area of a trapezoid is half the product of the length of its altitude and the sum of the lengths of its bases.

Theorem 7.5   If two triangles are similar, then the ratio of their areas equals the square of the ratio of the lengths of any two corresponding sides *or* the square of the ratio of the lengths of any two corresponding altitudes.

Theorem 7.6 $(C = 2\pi r)$ The circumference of a circle is equal to twice the product of π and the radius of the circle.

Theorem 7.7 $(A = \pi r^2)$ The area of a circle is the product of π and the square of the radius of the circle.

Theorem 8.1 $(PQ = \sqrt{(a - c)^2 + (b - d)^2})$ The distance between the points $P(a, b)$ and $Q(c, d)$ equals the square root of the sum of the squares of $(a - c)$ and $(b - d)$.

Theorem 8.2   An isometry preserves angle measure.

Theorem 8.3   Translations are isometries.

Theorem 8.4   Rotations are isometries.

Theorem 8.5   Reflections are isometries.

Theorem 9.1 $(S = \pi rs)$ The lateral surface area of a right circular cone is the product of π, the radius of the circle, and the distance from the vertex to any point on the circle.

Theorem 9.2 $(S = 2\pi ra)$ The lateral surface area of a right circular cylinder is two times the product of π, the radius of the circle, and the length of the altitude of the cylinder.

Theorem 9.3 $(S = 4\pi r^2)$ The surface area of a sphere is four times the product of π and the square of the radius.

Theorem 9.4 $(V = Ba)$ The volume of a prism or a circular cylinder equals the area of a base times the length of the altitude.

Theorem 9.5 $(V = Ba/3)$ The volume of a pyramid or a circular cone equals one third the area of the base times the length of the altitude.

Theorem 9.6 $(V = 4\pi r^3/3)$ The volume of a sphere equals four thirds the product of π and the cube of the radius.

Appendix D
List of Constructions

Construction 1.1   Construct a line segment equal in length to a given line segment.

Construction 1.2   Construct an angle equal in measure to a given angle.

Construction 1.3   Construct the perpendicular bisector of a given line segment.

Construction 1.4   Construct the bisector of a given angle.

Construction 1.5   Construct the perpendicular to a line from a point on the line.

Construction 1.6   Construct the perpendicular to a line from a point not on the line.

Construction 3.1   Through a point outside a line construct the parallel to the line.

Construction 3.2   Divide a line segment into a given number of congruent segments.

Construction 4.1   Construct the fourth proportional to three given line segments.

Construction 4.2   Construct a polygon similar to a given polygon.

Construction 4.3   Construct an equilateral triangle.

Construction 4.4   Construct a square.

Construction 6.1   Construct a circle containing three given points not all on one line.

Construction 6.2   Construct a tangent of a circle through a given point of the circle.

Construction 6.3   Construct a tangent of a circle through a given point in the exterior of the circle.

Construction 6.4   Construct a common external tangent of two given circles.

Construction 6.5   Construct a common internal tangent of two given circles.

Construction 6.6   Construct a segment whose length is the geometric mean between the lengths of two given segments.

Construction 7.1   Construct a triangle equal in area to a given polygon.

Construction 8.1   Construct the reflection of a point P with respect to a line k.

Appendix E
List of Symbols

{ }	set	m∠A	measure of angle A
∈	is an element of	△, △s	triangle(s)
∉	is not an element of	int△	interior of triangle
=	is equal to	ext△	exterior of triangle
≠	is not equal to	°	degree(s)
∅	empty set	π	pi (approximately 3.1416)
<	is less than	rad	radian(s)
>	is greater than	′	minute(s)
≤	is less than or equal to	″	second(s)
≥	is greater than or equal to	⊥	is perpendicular to
⊂	is a subset of	⌐	perpendicular rays (in a diagram)
⊄	is not a subset of	⟷	corresponds to
:	such that	↮	does not correspond to
∪	union	‖	is parallel to
∩	intersection	▱	parallelogram
|x|	the absolute value of x	▭	rectangle
PQ	the distance from P to Q	∼	is similar to
A−B−C	B is between A and C	$\sqrt{r}$	the square root of r
$\overline{AB}$	segment AB	⊙	circle
$\overleftrightarrow{AB}$	line AB	int⊙	interior of circle
$\overrightarrow{AB}$	ray AB	ext⊙	exterior of circle
≃	is congruent to	$\overset{\frown}{AB}$	arc AB
≇	is not congruent to	$m\overset{\frown}{AB}$	measure of arc AB
∠, ∠s	angle(s)	areaABCDE	area of polygon ABCDE
int∠	interior of angle	area⊙	area of circle
ext∠	exterior of angle		

Appendix F
List of Abbreviations

add.	addition		meas.	measure
adj.	adjacent		mult.	multiplication
alt.	altitude; alternate		opp.	opposite
approx.	approximately		post.	postulate
cm	centimeter(s)		pt., pts.	point, points
comp.	complementary		quad.	quadrilateral
corr.	corresponding		seg.	segment
cos	cosine		sin	sine
def.	definition		st.	straight
diag.	diagonal		state.	statement
ft	foot (feet)		supp.	supplementary
geom.	geometric		tan	tangent
hyp.	hypotenuse		trans.	transversal
in	inch(es)		vert.	vertical

Appendix G
Powers and Roots

Number	Square	Square root	Number	Square	Square root	Number	Square	Square root
1	1	1.000	36	1296	6.000	71	5041	8.426
2	4	1.414	37	1369	6.083	72	5184	8.485
3	9	1.732	38	1444	6.164	73	5329	8.544
4	16	2.000	39	1521	6.245	74	5476	8.602
5	25	2.236	40	1600	6.325	75	5625	8.660
6	36	2.449	41	1681	6.403	76	5776	8.718
7	49	2.646	42	1764	6.481	77	5929	8.775
8	64	2.828	43	1849	6.557	78	6084	8.832
9	81	3.000	44	1936	6.633	79	6241	8.888
10	100	3.162	45	2025	6.708	80	6400	8.944
11	121	3.317	46	2116	6.782	81	6561	9.000
12	144	3.464	47	2209	6.856	82	6724	9.055
13	169	3.606	48	2304	6.928	83	6889	9.110
14	196	3.742	49	2401	7.000	84	7056	9.165
15	225	3.873	50	2500	7.071	85	7225	9.220
16	256	4.000	51	2601	7.141	86	7396	9.274
17	289	4.123	52	2704	7.211	87	7569	9.327
18	324	4.243	53	2809	7.280	88	7744	9.381
19	361	4.359	54	2916	7.348	89	7921	9.434
20	400	4.472	55	3025	7.416	90	8100	9.487
21	441	4.583	56	3136	7.483	91	8281	9.539
22	484	4.690	57	3249	7.550	92	8464	9.592
23	529	4.796	58	3364	7.616	93	8649	9.644
24	576	4.899	59	3481	7.681	94	8836	9.695
25	625	5.000	60	3600	7.746	95	9025	9.747
26	676	5.099	61	3721	7.810	96	9216	9.798
27	729	5.196	62	3844	7.874	97	9409	9.849
28	784	5.292	63	3969	7.937	98	9604	9.899
29	841	5.385	64	4096	8.000	99	9801	9.950
30	900	5.477	65	4225	8.062	100	10000	10.000
31	961	5.568	66	4356	8.124			
32	1024	5.657	67	4489	8.185			
33	1089	5.745	68	4624	8.246			
34	1156	5.831	69	4761	8.307			
35	1225	5.916	70	4900	8.367			

Appendix H
Values of Trigonometric Functions

Degree Meas.	sin	cos	tan	Degree Meas.	sin	cos	tan	Degree Meas.	sin	cos	tan
0	0.0000	1.0000	0.0000	31	0.5150	0.8572	0.6009	61	0.8746	0.4848	1.8040
1	0.0175	0.9998	0.0175	32	0.5299	0.8480	0.6249	62	0.8829	0.4695	1.8807
2	0.0349	0.9994	0.0349	33	0.5446	0.8387	0.6494	63	0.8910	0.4540	1.9626
3	0.0523	0.9986	0.0524	34	0.5592	0.8290	0.6745	64	0.8988	0.4384	2.0503
4	0.0698	0.9976	0.0699	35	0.5736	0.8192	0.7002	65	0.9063	0.4226	2.1445
5	0.0872	0.9962	0.0875	36	0.5878	0.8090	0.7265	66	0.9135	0.4067	2.2460
6	0.1045	0.9945	0.1051	37	0.6018	0.7986	0.7536	67	0.9205	0.3907	2.3559
7	0.1219	0.9925	0.1228	38	0.6157	0.7880	0.7813	68	0.9272	0.3746	2.4751
8	0.1392	0.9903	0.1405	39	0.6293	0.7771	0.8098	69	0.9336	0.3584	2.6051
9	0.1564	0.9877	0.1584	40	0.6428	0.7660	0.8391	70	0.9397	0.3420	2.7475
10	0.1736	0.9848	0.1763	41	0.6561	0.7547	0.8693	71	0.9455	0.3256	2.9042
11	0.1908	0.9816	0.1944	42	0.6691	0.7431	0.9004	72	0.9511	0.3090	3.0777
12	0.2079	0.9781	0.2126	43	0.6820	0.7314	0.9325	73	0.9563	0.2924	3.2709
13	0.2250	0.9744	0.2309	44	0.6947	0.7193	0.9657	74	0.9613	0.2756	3.4874
14	0.2419	0.9703	0.2493	45	0.7071	0.7071	1.0000	75	0.9659	0.2588	3.7321
15	0.2588	0.9659	0.2679	46	0.7193	0.6947	1.0355	76	0.9703	0.2419	4.0108
16	0.2756	0.9613	0.2867	47	0.7314	0.6820	1.0724	77	0.9744	0.2250	4.3315
17	0.2924	0.9563	0.3057	48	0.7431	0.6691	1.1106	78	0.9781	0.2079	4.7046
18	0.3090	0.9511	0.3249	49	0.7547	0.6561	1.1504	79	0.9816	0.1908	5.1446
19	0.3256	0.9455	0.3443	50	0.7660	0.6428	1.1918	80	0.9848	0.1736	5.6713
20	0.3420	0.9397	0.3640	51	0.7771	0.6293	1.2349	81	0.9877	0.1564	6.3138
21	0.3584	0.9336	0.3839	52	0.7880	0.6157	1.2799	82	0.9903	0.1392	7.1154
22	0.3746	0.9272	0.4040	53	0.7986	0.6018	1.3270	83	0.9925	0.1219	8.1443
23	0.3907	0.9205	0.4245	54	0.8090	0.5878	1.3764	84	0.9945	0.1045	9.5144
24	0.4067	0.9135	0.4452	55	0.8192	0.5736	1.4281	85	0.9962	0.0872	11.430
25	0.4226	0.9063	0.4663	56	0.8290	0.5592	1.4826	86	0.9976	0.0698	14.301
26	0.4384	0.8988	0.4877	57	0.8387	0.5446	1.5399	87	0.9986	0.0523	19.081
27	0.4540	0.8910	0.5095	58	0.8480	0.5299	1.6003	88	0.9994	0.0349	28.636
28	0.4695	0.8829	0.5317	59	0.8572	0.5150	1.6643	89	0.9998	0.0175	57.290
29	0.4848	0.8746	0.5543	60	0.8660	0.5000	1.7321	90	1.0000	0.0000	—
30	0.5000	0.8660	0.5774								

Answers to
Odd-Numbered Exercises

Exercise Set 1.1

1. Infinite
3. Infinite
5. Finite
7. Answers may vary: $\{x: x < 3 \text{ and } x \geq 4\}$, $\{x: x \in A \cap B \text{ and } x \notin A \cap B\}$, $\{\text{oranges with teeth}\}$
9. $\subset$
11. $\subset$
13. $\subset$
15. $\subset$
17. B
19. D
21. F
23. $\in$
25. $\notin$
27. $\in$
29. $\in$
31. $\subset$
33. $\not\subset$
35. A
37. $\{99, 108, 111\}$
39. $\notin$
41. $\in$
43. 121
45. 38
47. 20
49. 1078

Exercise Set 1.2

1. Postulate 1.5
3. Postulate 1.2
5. Postulate 1.4
7. 7
9. 0
11. 52
13. d if $d \geq 0$, $-d$ if $d < 0$

15. x^2
17. $x - 3$
19. $a - b$
21. $-x - y$
23. 2
25. 5
27. 1/2
29. 13/2
31. $|q - p| = |p - q|$
33. 11 parsecs, 16 parsecs
35. Four points do not necessarily determine a unique plane.

Exercise Set 1.3

1. $\overline{AE}, \overline{CE}, \overline{AC}, \overline{AF}, \overline{BF}, \overline{AB}, \overline{AD}, \overline{CD}$
3. $\vec{AE}, \vec{AB}, \vec{AF}, \vec{AC}, \vec{AD}, \vec{BF} = \vec{BA}, \vec{FB} = \vec{FA}, \vec{EC} = \vec{EA}, \vec{CE} = \vec{CA}$
5. $=, \simeq$
7. $\simeq, \neq$
9. $\neq, \not\simeq$
11. C, F, G
13. Same side
15. Same side
17. D
19. None
21. $\angle BCD, \angle CBE, \angle ABF$
23. Ray AB is the union of line segment AB with the set of all points P such that B is between A and P.
25. Answers may vary.
27. Answers may vary.

Exercise Set 1.4

1. $80°51'$
3. $62°13'12''$
11. $90°$
13. $22°30'$

15. $\angle ABC = \angle DEF$ means that the two angles (sets of points) are identical. $\angle ABC \simeq \angle DEF$ means that the two angles have the same measure.

17. Approximately 150°

19. 6°

21. Approximately 180°

Exercise Set 1.5

1. 60°

3. 23″ 38′

5. 89° 56′ 1″

7. $\pi/4$ rad

9. 30°

11. 145°

13. 104° 24′ 42″

15. $\pi/2$ rad

17. 45°, 90°

19. 45°

21. F $\notin$ int$\angle BAE$

23. Obtuse

25. Right

27. Acute

29. Right

31. Acute

33. Not an angle (a ray)

35. Answers may vary.

Chapter 1 Review Exercises

1. Always

3. Sometimes

5. Always

7. Sometimes

9. Sometimes

11. Infinite

13. Intersection (or subset)

15. True

17. B

19. Supplement

21. Infinite if $A \neq B$; $\emptyset$ if $A = B$

23. Finite

25. $\emptyset$

27. $\emptyset$

29. A, B, D

31. $\{a, b, c, d, e, 4\}$

33. $\emptyset$

35. $\{1, 4\}$

37. 3

39. x^2

41. $b - a$

43. $\angle 6, \angle 7; \angle 6, \angle BAC; \angle 7, \angle BAC$

45. All pairs

47. 10°

49. None

51. 100°

53. $\pi/3$ rad

Exercise Set 2.1

1. I will not earn enough money.

3. You will turn the television off.

5. Two angles are complementary.

7. $AB \neq CD$

9. A point is not in the interior of $\triangle PQR$.

11. False

13. False

15. False

17. True

19. True

21. I will earn enough money and I will buy a car.

23. Two angles are not complementary and two angles are adjacent.

25. A point is in the interior of $\triangle PQR$ and a point is in int$\angle Q$.

27. You will not turn the television off or you will not be able to study better.

29. $\overline{AB} \simeq \overline{CD}$ or m$\angle A$ = m$\angle B$.

31. True

33. False

35. False

37. True

39. False

45.

p	$not\ p$	$p\ or\ not\ p$
T	F	T
F	T	T

47.

p	q	r	$not\ q$	$r\ or$ $not\ q)$	$p\ and$ $(r\ or$ $not\ q)$
T	T	T	F	T	T
T	T	F	F	F	F
T	F	T	T	T	T
T	F	F	T	T	T
F	T	T	F	T	F
F	T	F	F	F	F
F	F	T	T	T	F
F	F	F	T	T	F

49.

p	r	not p	(p or r)	not p and (p or r)
T	T	F	T	F
T	F	F	T	F
F	T	T	T	T
F	F	T	F	F

51.

p	q	not p	not q	p or q	not (p or q)	not p and not q
T	T	F	F	T	F	F
T	F	F	T	T	F	F
F	T	T	F	T	F	F
F	F	T	T	F	T	T

53. $S \cup T = \{2, 4, 8, 10\}, (S \cup T)' = \{6\}$.
$S' = \{6, 10\}, T' = \{4, 6, 8\}, S' \cap T' = \{6\}$.

55. $S \cup T = \{$the set of non-zero integers$\}, (S \cup T)' = \{0\}$.
$S' = \{$the set of non-positive integers$\}, T' = \{$the set of non-negative integers$\}, S' \cap T' = \{0\}$.

Exercise Set 2.2

1. I earn enough money. I will buy a car.
3. You turn the television off. You will be able to study better.
5. $A \simeq B$. m∠A = m∠B.
7. It is red in the morning. Sailors take warning.
9. A statement is a theorem. The statement always has an hypothesis and a conclusion.
11. If I will buy a car, then I earn enough money.
13. If you will be able to study better, then you turn the television off.
15. If m∠A = m∠B, then $A \simeq B$.
17. If sailors take warning, then it is red in the morning.
19. If a statement always has a hypothesis and a conclusion, then the statement is a theorem.
21. If I do not earn enough money, then I will not buy a car.
23. If you do not turn the television off, you will not be able to study better.
25. If $A \not\simeq B$, then m∠$A \neq$ m∠B.
27. If it is not red in the morning, sailors do not take warning.
29. If a statement is not a theorem, then it does not always have a hypothesis and a conclusion.
31. If I will not buy a car, then I do not earn enough money.

33. If you will not be able to study better, then you do not turn the television off.
35. If m∠$A \neq$ m∠B, then $A \not\simeq$ B.
37. If sailors do not take warning, then it is not red in the morning.
39. If a statement does not always have a hypothesis and conclusion, then the statement is not a theorem.
41. You are on the moon. The acceleration due to gravity is only about one-sixth that on earth. If the acceleration due to gravity is only about one-sixth that on earth, then you are on the moon. If you are not on the moon, then it is not true that the acceleration due to gravity is only about one-sixth that on earth. If it is not true that the acceleration due to gravity is only about one-sixth that on earth, then you are not on the moon.
43. Rubber is warmed. Rubber has the property of becoming sticky. If rubber has the property of becoming sticky, then it is warmed. If rubber is not warmed, then it does not have the property of becoming sticky. If rubber does not have the property of becoming sticky, then it is not warmed.
45. A reactive metal is made the anode of an electrolytic cell. Anodic oxidation may involve oxidation of the metal composing the electrode. If anodic oxidation may involve oxidation of the metal composing the electrode, then a reactive metal is made the anode of an electrolytic cell. If a reactive metal is not made the anode of an electrolytic cell, then it is not true that the anodic oxidation may involve oxidation of the metal composing the electrode. If it is not true that anodic oxidation may involve oxidation of the metal composing the electrode, then a reactive metal is not made the anode of an electrolytic cell.
47. Any zygote is formed by combination with a normal gamete. Any zygote will be aneuploid and lethal. If any zygote will be aneuploid and lethal, then it is formed by combination with a normal gamete. If any zygote is not formed by combination with a normal gamete, then it will not be aneuploid and lethal. If any zygote will not be aneuploid and lethal, then it is not formed by combination with a normal gamete.
49. A taxpayer uses the accrual basis. A taxpayer reports income when it is earned, even though not yet received, and deducts expenses when they are incurred, even though not yet paid. A taxpayer who reports income when it is earned, even though not

yet received, and deducts expenses when they are incurred, even though not yet paid, uses the accrual basis. If a taxpayer does not use the accrual basis then it is not true that the taxpayer reports income when it is earned, even though not yet received, and deducts expenses when they are incurred, even though not yet paid. If it is not true that a taxpayer reports income when it is earned, even though not yet received, and deducts expenses when they are incurred, even though not yet paid, then the taxpayer does not use the accrual basis.

51. Answers may vary.
53. Answers may vary.

Exercise Set 2.3

1. Given: $\triangle ABC$
Prove: $m\angle A + m\angle B + m\angle C = 180°$
3. Given: $\angle B$ is a right $\angle$
Prove: $(AC)^2 = (AB)^2 + (BC)^2$
5. Given: Intersecting line j and k,
$m\angle 1 = m\angle 2, m\angle 3 = m\angle 4$
Prove: $\overrightarrow{BA} \perp \overrightarrow{BC}$
7. Given: $AF = BF, BE = CE, AD = CD,$
$j \perp \overline{AB}, i \perp \overline{BC}, k \perp \overline{AC}$
Prove: $i, j,$ and k are concurrent
9. Given: $m\angle 1 + m\angle 2 = 180°$
$m\angle 1 > 90°$
Prove: $m\angle 2 < 90°$
11. Given: $x + 7 = 9$
Prove: $x = 2$
13. Given: $f(x) = x - 9$ and $g(y) = y + 6$
Prove: $f(x) + g(y) = x + y - 3$

Exercise Set 2.4

1. Statement to be proved, diagram, given, prove, proof
3. $\angle 1, \angle 3; \angle 2, \angle 4; \angle 5, \angle 7; \angle 6, \angle 8; \angle 9, \angle 11; \angle 10, \angle 12; \angle 13, \angle 15; \angle 14, \angle 16$
5. 1. $A-B-C, B-C-D, AB = CD$
2. Def. of between
3. Substitution
4. Commutativity
5. Transitivity
6. Def. of between and substitution

7. 1. $a, b,$ and c are lines, $m\angle 1 + m\angle 3 = 180°$
2. Def. of supp.
3. Substitution
4. $m\angle 2 = m\angle 3$
5. Vert. $\angle$s are $\simeq$
6. $m\angle 2 = m\angle 4$
9. 1. $\angle ABC$ is a right $\angle$, $\angle EFG$ and $\angle DBC$ are comp.
2. Def. of right $\angle$
3. Angle Add. Post.
4. $m\angle ABD + m\angle DBC = 90°$
5. Def. of comp.
6. Comps. of $\simeq \angle$s are $\simeq$
11. 1. $\angle ABC$ and $\angle ABD$ are supp., $\overrightarrow{BE}$ bis. $\angle ABC$, $\overrightarrow{BF}$ bis. $\angle ABD$
2. Def. of $\angle$ bisector
3. Angle Add. Post.
4. Substitution
5. Add. Axiom of =
6. Def. of supp.
7. Substitution
8. Mult. Axiom of =
9. Def. of $\perp$ and meas.
13. Given: $\angle A$
Prove: $\angle A \simeq \angle A$
1. $\angle A$
2. Reflexivity
3. $\angle A \simeq \angle A$
15. Given: $\angle A \simeq \angle B, \angle B \simeq \angle C$
Prove: $\angle A \simeq \angle C$
Proof:
1. $\angle A \simeq \angle B$, $\angle B \simeq \angle C$ — 1. Given
2. $m\angle A = m\angle B$, $m\angle B = m\angle C$ — 2. Def. of $\simeq \angle$s
3. $m\angle A = m\angle C$ — 3. Transitivity
4. $\angle A \simeq \angle C$ — 4. Def. of $\simeq \angle$s

Exercise Set 2.5

1. $A \longleftrightarrow F, C \longleftrightarrow E, B \longleftrightarrow D$
3. $\angle A \longleftrightarrow \angle F, \angle C \longleftrightarrow \angle E, \angle B \longleftrightarrow \angle D$
5. $\triangle KLM \simeq \triangle TSR$
7. $\angle K \longleftrightarrow \angle R$
9. $\angle KIJ \longleftrightarrow \angle RPQ$ and def. of meas.
11. None
13. None
15. None

17. $\triangle ABC \simeq \triangle FDE$, SSS = SSS
19. $\triangle ABC \simeq \triangle FED$, ASA = ASA
21. Show $\triangle ABD \simeq \triangle CDB$ by SSS = SSS and use corr. parts of $\simeq \triangle$s are $\simeq$.
23. Show $\triangle BCE \simeq \triangle BDE$ by ASA = ASA. Thus, $BC = BD$ by corr. parts of $\simeq \triangle$s are $\simeq$. Thus, $\triangle ABC \simeq \triangle ABD$ by SSS = SSS.
25. Show m$\angle BAC = 61°$. Show $AC = EF$. Use ASA = ASA.
27. SSS = SSS

Exercise Set 2.6

(In all the following answers, theorems are referred to by numbers, but the student should quote the entire theorem.)

1. No. The third side may not be congruent to the two given congruent sides.
3. $39°$
5. $60°$, $60°$
7. Insufficient information
9. $\angle 2 \simeq \angle 3$ by Theorem 2.4. Use Theorem 2.1.
11. $\angle 2 \simeq \angle B$ since corr. parts of $\simeq \triangle$s are $\simeq$. Therefore, $\angle 1 \simeq \angle B$. Use Theorem 2.5.
13. Use Theorem 2.4 to get $\angle Q \simeq \angle R$. Use ASA = ASA.
15. $\triangle ACE \simeq \triangle BDE$ by SAS = SAS. Use corr. parts of $\simeq \triangle$s are $\simeq$.
17. Use the Angle Add. Post. Show that $AC = BC$, so that $\triangle DAC \simeq \triangle DBC$ by SSS = SSS. Use corr. parts of $\simeq \triangle$s are $\simeq$.
19. Use Theorem 2.4 twice.
21. Use Theorem 2.5 twice.

Exercise Set 2.7

1. Both are line segments with one endpoint at a vertex point, but an altitude is perpendicular to the opposite side and the other endpoint of the median is the midpoint of the opposite side.
7. Yes
9. If $AB = CD$, then m$\angle 1 = $ m$\angle 2$.
11. Use SAS = SAS.
13. Use Theorem 2.6 to show that m$\angle 6 > $ m$\angle 3$ and m$\angle 3 > $ m$\angle 1$.
15. Show that $\triangle CAE \simeq \triangle CBD$ by using ASA = ASA. Then, $CD = CE = AD = BE$.

17. Show that $\triangle ACD \simeq \triangle BCE$ by SAS = SAS and thus, $\angle DAC \simeq \angle EBC$. It follows that $\angle CAB \simeq \angle CBA$ and thus, $AC = BC$ by Theorem 2.5.
19. See the suggestion after Theorem 2.7.
21. Answers may vary.

Chapter 2 Review Exercises

1. Always
3. Sometimes
5. Sometimes
7. Sometimes
9. Always
11. Does not include
13. Congruent
15. Sides
17. True
19. Greater
21. It is raining.
23. Triangles that are not scalene are isosceles.
25. The sugar is not sweet, when the beets are not a treat.
27. $\angle 1$, $\angle 2$; $\angle 1$, $\angle 4$; $\angle 2$, $\angle 3$; $\angle 3$, $\angle 4$; $\angle 5$, $\angle 6$; $\angle 5$, $\angle 8$; $\angle 6$, $\angle 7$; $\angle 7$, $\angle 8$
29. 10
31. $68°$
33. ASA = ASA
35. SSS = SSS
37. $AB = BC = 23$
39. m$\angle ACD < 72°$
41. *Proof:*

1. $AD = BE$	1. Given
2. $DE = DE$	2. Reflexivity
3. $AD + DE =$ $BE + ED$	3. Add. Axiom of $=$
4. $AE = BD$	4. Def. of between and substitution

43. Use Theorem 2.4 and Theorem 2.5.
47. Show that $\triangle ABD \simeq \triangle ACD$ by using SSS = SSS. Thus, $\angle BAD \simeq \angle CAD$. Therefore, by def., $\overrightarrow{AD}$ bisects $\angle BAC$.
49. *Proof:*

1. $A-D-B$, $C-E-D$, $AC = BC$, $AE = BE$	1. Given
2. $CE = CE$	2. Reflexivity
3. $\triangle AEC \simeq \triangle BEC$	3. SSS = SSS
4. $\angle CEA \simeq \angle CEB$	4. Corr. parts of $\simeq$ $\triangle$s are $\simeq$

5. $\angle AED \simeq \angle BED$
6. $\angle EAD \simeq \angle EBD$
7. $\triangle ADE \simeq \triangle BDE$

8. $\angle EDA \simeq \angle BDA$

9. $\overline{CD} \perp \overline{AB}$
10. $\overline{AD} \simeq \overline{BD}$

11. $\overline{CD}$ is a median

5. Theorem 2.1
6. Theorem 2.4
7. ASA = ASA
   (Statements 1, 5, 6)
8. Corr. parts of $\simeq$
   $\triangle$s are $\simeq$
9. Def. of $\perp$
10. Corr. parts of $\simeq$
    $\triangle$s are $\simeq$
11. Def. of median

51. *Proof:*

1. $A-D-B$,
   $m\angle 1 = m\angle 2$,
   $\overline{CD}$ is an altitude
2. $\angle ADC \simeq \angle BDC$
3. $CD = CD$
4. $\triangle ADC \simeq \triangle BDC$
5. $\overline{AD} \simeq \overline{BD}$

6. $\overline{CD}$ is a median

1. Given

2. Def. of alt. and $\perp$
3. Reflexivity
4. ASA = ASA
5. Corr. parts of $\simeq$
   $\triangle$s are $\simeq$
6. Def. of median

5. Yes. Theorem 3.4. Yes; def. of ‖ line segments.
7. No. By def. of triangle, each side intersects each of the other sides. Therefore, by def. they cannot be ‖.
9. Use Theorem 3.1.
11. Use Theorem 3.2.
13. Use Theorem 3.1.
15. Use Theorem 3.3.
17. Either $m\angle 4 = m\angle 5$ or $m\angle 4 \neq m\angle 5$. Assume $m\angle 4 = m\angle 5$. Then $BC = BD$ by Theorem 2.5 and $m\angle 3 = m\angle 6$ by Theorem 2.1. We are given that $AC = DE$. Therefore, $\triangle ABC \simeq \triangle EBD$ by SAS = SAS. Contradiction. Therefore, $m\angle 4 \neq m\angle 5$.
19. Either $AB = BE$ or $AB \neq BE$. Assume $AB = BE$. Then by Theorem 2.4, $m\angle A = m\angle E$. Since $AC = DE$ is given, $\triangle ABC \simeq \triangle EBD$ by SAS = SAS. Since corr. parts of $\simeq$ $\triangle$s are $\simeq$, $m\angle 3 = m\angle 6$. Therefore, $m\angle 4 = m\angle 5$ since supp. of $\simeq$ $\angle$s are $\simeq$. Contradiction. Therefore, $AB \neq BE$.

Exercise Set 3.1

1. Yes. Two distinct line segments in a plane are parallel if the lines containing them are parallel.
3. No. $\overleftrightarrow{AE}$ is not parallel to $\overleftrightarrow{CD}$, since through A there exists exactly one parallel to $\overleftrightarrow{CD}$ and that parallel is given as $\overleftrightarrow{AB}$, but $\overleftrightarrow{AE} \neq \overleftrightarrow{AB}$. Hence, the line segments contained in $\overleftrightarrow{AE}$ and $\overleftrightarrow{CD}$ are not parallel.
5. No. If there existed such a Q, both $\overleftrightarrow{EQ}$ and $\overleftrightarrow{CD}$ would be lines through Q parallel to $\overleftrightarrow{AB}$. This is impossible by the Parallel Postulate.
7. No. If $j \parallel m$, then k would equal m by the Parallel Postulate.
9. No. If $P \in j \cap m$ and $p \in j \cap k$, then we would have $P \in k \cap m$.

Exercise Set 3.2

1. $AB < 8$, $BC > 8$; Theorem 3.3.
3. $m\angle C < 42°$; Theorem 3.2, $m\angle B > 96°$ since $m\angle B + m\angle C = 138°$ and $m\angle C < 42°$.

Exercise Set 3.3

1. $\angle 2$, $\angle 4$, $\angle 5$, $\angle 7$
3. $\angle 2$, $\angle 7$; $\angle 4$, $\angle 5$
5. $\angle 1$, $\angle 5$; $\angle 2$, $\angle 6$; $\angle 3$, $\angle 7$; $\angle 4$, $\angle 8$
7. $\angle 1$, $\angle 8$; $\angle 2$, $\angle 7$; $\angle 1$, $\angle 12$; $\angle 3$, $\angle 10$; $\angle 5$, $\angle 10$; $\angle 7$, $\angle 9$
9. $m\angle 4 = m\angle 5 = m\angle 8 = 137°$, $m\angle 2 = m\angle 3 = m\angle 6 = m\angle 7 = 43°$
11. Show that $m\angle 1 = m\angle 5$, $m\angle 3 = m\angle 6$, and use Add. Axiom of =.
13. Show that $m\angle 1 = m\angle 5$. Then use Theorem 2.6 and substitution.
15. Use Theorem 3.8 and the Angle Addition Postulate.
17. See Exercise 16 and also show that $m\angle BAC = m\angle CED$. Then use Theorem 2.1.
19. Show that $m\angle GBA = m\angle BAD$. Hence, $m\angle BAD = m\angle EDC$. Use Theorem 3.8.
21. Show that $m\angle IAB = m\angle ABE$ and hence, $\angle ABE$ is supp. to $\angle BED$. Use Theorem 3.10.
23. Assume $\triangle ABC \simeq \triangle DEF$. Then by Exercise 22, $\overleftrightarrow{AB} \parallel \overleftrightarrow{DE}$. Contradiction. Therefore, $\triangle ABC \neq \triangle DEF$.

25. Let $BC \cap EF = \{P\}$. By Theorem 3.7, $m\angle ABC = m\angle BPE$. By Theorem 3.10, $\angle BPE$ and $\angle E$ are supp. Use substitution and def. of supp.

27. (if) Given $\triangle ABC \simeq \triangle CDA$, show that $AB = CD, BC = DA, \angle BAC \simeq \angle DCA$, and $\angle BCA \simeq \angle DAC$. Use the Angle Add. Post. to show that $\angle DAB \simeq \angle BCD$. Then use SAS = SAS to prove that $\triangle ABD \simeq \triangle CDB$.

(only if) Given $\triangle ABD \simeq \triangle CDB$, show that $AB = CD, BC = DA, \angle ABD \simeq \angle CDB$, and $\angle ADB \simeq \angle CBD$. Use the Angle Add. Post to show that $\angle ABC \simeq \angle CDA$. Then use SAS = SAS to prove that $\triangle ABC \simeq \triangle CDA$.

29. Assume $m\angle ABC > m\angle DCB$. Then there is a ray $\overrightarrow{BF}$ such that $m\angle DCB = m\angle CBF$. By Theorem 3.6, $\overleftrightarrow{BF} \| \overleftrightarrow{CD}$. But by hypothesis $\overleftrightarrow{AB} \| \overleftrightarrow{CD}$. This contradicts the Parallel Post.

31. Use a proof similar to that of Theorem 3.8, but using an $\angle 4$, where $\angle 1$ and $\angle 4$ are alt. ext. $\angle$s.

33. $m\angle B = 70°$, $m\angle C = 35°$

35. 45°

37. 135°

39. $m\angle A = 30°$, $m\angle ABF = 30°$, $m\angle DBC = 30°$, $m\angle C = 15°$.

43. Show that $m\angle CDB = 110°$ by using Theorem 3.11. Use Theorem 3.8.

45. Show that $m\angle ACB = m\angle ECD$, and use Theorem 3.11 to show that $m\angle B = m\angle D$. Then use ASA = ASA. (See Exercise 46).

47. Let P be a point such that $A - B - P$. Show that $m\angle A = m\angle DBP$ and that $m\angle C = m\angle CBD$. Use Add. Axiom of = and Angle Add. Post.

Exercise Set 3.4

1. 0

3. 5

5. 180°

7. 720°

9. 9

11. 18

13. $1720/180 = 86/9$ is not an integer

15. Let Column A = sum of measures of int. $\angle$s,
Column B = measure of each int$\angle$,
Column C = measure of each ext$\angle$,
Column D = sum of measure of ext. $\angle$s.
(All measures are in degrees.)

n	A	B	C	D
3	180	60	120	360
4	360	90	90	360
5	540	108	72	360
6	720	120	60	360
7	900	128 4/7	51 3/7	360
8	1080	135	45	360
9	1260	140	40	360
10	1440	144	36	360
k	$(k-2)180$	$180-360/k$	$360/k$	360

17. Two triangles are formed by a diagonal and triangles are stable. A rope can sag, causing a change in the length of the diagonal.

Exercise Set 3.5

1. $m\angle C = 50°$, $m\angle B = m\angle D = 130°$

3. Yes, by def., all squares are rectangles.

5. Yes, by def., a square is a rhombus that is a rectangle.

7. Theorem 3.15

9. Corollary to Theorem 3.15

21. If $AB = CD$ and $AD = BC$ in quad. $ABCD$, then $\triangle ACD \simeq \triangle CAB$. Thus, $\angle ACD \simeq \angle CAB$ and $\angle DAC \simeq \angle BCA$. Therefore, $\overline{AB} \| \overline{CD}$ and $\overline{AD} \| \overline{BC}$. If $\angle A \simeq \angle C$ and $\angle B \simeq \angle D$ in quad. $ABCD$, then $2m\angle A + 2m\angle B = 360°$ by Theorem 3.13. Thus $m\angle A + m\angle B = 180°$, and by substitution $m\angle A + m\angle D = 180°$. Therefore, by Theorem 3.10, $\overline{AD} \| \overline{BC}$ and $\overline{AB} \| \overline{CD}$.

23. (a) If a quadrilateral is a parallelogram, then its diagonals bisect each other. (b) If the diagonals of a quadrilateral bisect each other, then the quadrilateral is a parallelogram.

Proof:

(a) In $\square ABCD$, let $\overline{AC} \cap \overline{BD} = \{E\}$. Show that $\triangle ABE \simeq \triangle CDE$ by ASA = ASA and, therefore, $AE = CE$ and $BE = DE$.

(b) In quad. $ABCD$, let $\overline{AC} \cap \overline{BD} = \{E\}$. Show that $\triangle ABE \simeq \triangle CDE$ by SAS = SAS and, hence, $\angle EAB \simeq \angle ECD$. Likewise, $\triangle AED \simeq \triangle CEB$ and, hence, $\angle DAE \simeq \angle BCE$. Therefore, $\overline{AB} \| \overline{CD}$ and $\overline{AD} \| \overline{BC}$.

Exercise Set 3.6

1. 3
3. $m\angle A = m\angle B = 80°$, $m\angle D = 100°$
5. 5
7. $45°$
9. Theorem 3.22
13. By Theorem 3.22, $\overline{MQ} \parallel \overline{BD}$, $\overline{PN} \parallel \overline{BD}$, $MQ = BD/2$, and $PN = BD/2$. By Theorem 3.20, $\overline{MQ} \parallel \overline{PN}$. By substitution, $MQ = PN$. Now see Exercise Set 3.5, No. 20.
15. Show that $\triangle CDF \simeq \triangle BGF$ and thus, $AD = BG$ and $\angle G \simeq \angle CDF$. Hence, $\overline{AD} \parallel \overline{BG}$ and by Exercise 3.5, No. 20, quad. $ABGD$ is a parallelogram. Therefore, $\overline{AB} \parallel \overline{DG}$ and $AB = DG$. Since $DG = 2DF$, we have $AB = 2DF$ and thus, $DF = AB/2$.
17. By def. of median $AE = DE$. By Theorem 3.7, $\angle CDE \simeq \angle PAE$. Since vert. $\angle$s are $\simeq$, $\angle PEA \simeq \angle CED$. Thus, $\triangle APE \simeq \triangle DCE$ by ASA = ASA. Since corr. parts of $\simeq \triangle$s are $\simeq$, $CE = PE$. But by def. of median $CF = FB$. Therefore, by Theorem 3.22, $EF = PB/2$. But $PB = PA + AB$ and $PA = CD$, so that $EF = (CD + AB)/2$.

Chapter 3 Review Exercises

1. Always
3. Sometimes
5. Always
7. Always
9. Never
11. Supplementary
13. True
15. True
17. Octagon
19. True
21. $60°$, $120°$
23. $50°$ or $130°$. See Exercise 3.3, No. 24 and 25.
25. $72°$
27. $m\angle A = 60°$, $m\angle B = 84°$
29. 12
31. 4
33. $120°$
35. 24
37. By Theorem 3.18, the diagonals of a parallelogram bisect each other. Hence, in square $ABCD$, if $\overline{AC} \cap \overline{BD} = \{E\}$, then $BE = ED$ and $AE = EC$. Show that $\triangle AEB \simeq \triangle AED$ by SSS = SSS, and thus, $\angle AEB \simeq \angle AED$. Use the def. of $\perp$.

39. Show that $\triangle ABD \simeq \triangle CBD$ by SSS = SSS. Hence, $\angle CBD \simeq \angle ABD$. If $\overline{AC} \cap \overline{BD} = \{E\}$, show that $\triangle ABE \simeq \triangle CBE$ by SAS = SAS, and use the def. of $\perp$.
41. Show that $AC = BD$ by using the def. of between and the Add. Axiom of =. Then show that $\triangle ACE \simeq \triangle DBF$ by SAS = SAS. Hence, $\angle EAC \simeq \angle FDB$. Use Theorem 3.6.
43. If $m\angle A = m\angle B + m\angle C$ and $m\angle A + m\angle B + m\angle C = 180°$, then $2m\angle A = 180°$, so $m\angle A = 90°$.
45. $\overline{AD} \parallel \overline{BE}$ by Theorem 3.4. Show that $\angle DAP \simeq \angle PBE$ and that $\angle APD \simeq \angle BPE$, so that $\triangle APD \simeq \triangle BPE$ by ASA = ASA. Hence, $AD = BE$. Now use Exercise Set 3.5, No. 20.
47. Use Theorem 3.22 repeatedly and then use SSS = SSS repeatedly.

Exercise Set 4.1

1. If the means of a proportion are exchanged, the resulting expression is a proportion. Given $a/b = c/d$, using the mult. axiom of =, multiply both sides by bd, yielding $ad = bc$. Then using the mult. axiom of =, multiply both sides by $1/cd$. The result follows.
3. In a proportion, the geometric mean equals plus or minus the square root of the product of the extremes. Given $a/b = b/c$, using the mult. axiom of =, multiply both sides by bc, yielding $b^2 = ac$. The result follows.
5. If in a proportion each denominator is subtracted from the corresponding numerator, the new ratios form a proportion. Given $a/b = c/d$, using the addition axiom of =, add -1 to both sides, yielding $a/b - 1 = c/d - 1$. The left side equals $(a - b)/b$ and the right side equals $(c - d)/d$, thus, the result follows.
7. $2/3$
9. $3/1$
11. $1/6$
13. $3/4$
15. 4
17. 22
19. $8/27$
21. 6
23. $32/5$
25. ± 9
27. $\pm 3/2$
29. $3/9 = 4/12 = 5/15$
31. $4/6 = 8/12 = 12/18 = 16/24$

33. 40/11 qts of water, 4/55 cup of salt, 4/11 lb of carrots, 48/55 of a chicken, 12/55 lb celery

35. 40/11 cm, 1200/11 cm

Exercise Set 4.2

1. $\angle A \simeq \angle R$, $\angle B \simeq \angle S$, $\angle C \simeq \angle Q$; $AB/RS = AC/RQ = BC/SQ$

3. $\angle QKL \simeq \angle QKA$, $\angle L \simeq \angle A$, $\angle M \simeq \angle B$, $\angle N \simeq \angle C$, $\angle P \simeq \angle D$, $\angle PQK \simeq \angle DQK$; $KL/AK = LM/AB = MN/BC = NP/CD = PQ/DQ = QK/QK = 1$

5. 160 ft

7. 500/7 ft

9. 7/3 in, 21/17 in

11. 24 ft by 20 ft

13. $\triangle ABC \sim \triangle FED$

15. $\triangle PQR \sim \triangle STR$

17. 8/3

19. 2/7

21. 16/3

29. Use Theorem 2.4 to show $\angle CDE \simeq \angle CED$ and $\angle CAB \simeq \angle CBA$. Then, show that $\angle CDE \simeq \angle CAB$ and $\angle CED \simeq \angle CBA$, by Theorem 3.11.

31. Use AA ~ AA.

33. Use AA ~ AA.

35. Use AA ~ AA.

37. Use Theorem 3.11 and AAA ~ AAA

Exercise Set 4.3

1. SAS ~ SAS, $\triangle ABC \sim \triangle EDC$

3. No

5. AA ~ AA, $\triangle VWX \sim \triangle ZYW$

7. 6/5

9. 10/7

11. 16/3

13. Use AA ~ AA on the $\triangle$s formed by the altitudes.

15. Use AA ~ AA on the $\triangle$s formed by the bisectors.

17. Use AA ~ AA.

19. See the hint provided.

Exercise Set 4.4

1. $\overline{PA}$, $\overline{PB}$, $\overline{PC}$, $\overline{PD}$, $\overline{PE}$, $\overline{PF}$

5. $\overline{PA}$, $\overline{PB}$, $\overline{PC}$, $\overline{PD}$

7. $\overline{PE}$, $\overline{PF}$, $\overline{PG}$, $\overline{PH}$

9. 90°

11. 60°

13. 45°

15. 36°

17. 6

19. 15

21. 72

23. 17

25. $144n$

37. 51

39. 18

41. 233.6

43. Look at the $\triangle$s formed by two radii and a side and use SSS = SSS.

45. Both have measures of 72°. (See Exercise 39)

Chapter 4 Review Exercises

1. Never

3. Always

5. Always

7. Never

9. Never

11. Proportion, ratios

13. True

15. Congruent

17. Center, vertices or sides

19. May

21. 3/2

23. 1/704

25. 15/7

27. 1/4

29. 6

31. ±13

33. $\angle A \simeq \angle D$, $\angle C \simeq \angle E$, $\angle B \simeq \angle F$; $AC/DE = AB/DF = CB/EF$

35. 480 ft

37. 28/5

39. 2

41. 30°

43. 18/5

45. 6

47. 6

49. 3/8

51. Use def. of similar.

53. Show $\triangle ABC \sim \triangle DBE$ by SAS ~ SAS. Then use transitivity with Exercise 52.

55. Let $\overline{QU} \parallel \overline{RT}$ as shown. Then $m\angle SRT = m\angle QRT$ by def. of bisector, $m\angle SRT = m\angle RUQ$ by

Theorem 3.8, and $m\angle QRT = m\angle RQU$ by Theorem 3.7. Thus, $m\angle RUQ = m\angle RQU$. Therefore, by Theorem 2.5, $UR = QR$. Since, $\overline{QU} \parallel \overline{RT}$, it follows that $PT/QT = PR/UR$. Substituting yields $PT/QT = PR/QR$.

Exercise Set 5.1

1. Yes
3. Yes
5. No
7. Yes
9. Yes
11. $2\sqrt{23}$
13. $17\sqrt{3}$
15. Does not simplify
17. $15\sqrt{105}$
19. $\sqrt{7}/7$
21. $\sqrt{13}/13$
23. $\sqrt{6}/2$
25. $34\sqrt{53}/53$
27. $13\sqrt{5730}/450$
29. $\pm\sqrt{7}$
31. -2
33. $-3, 7$
35. $(-3 \pm \sqrt{73})/4$
37. ± 6
39. $\pm 3\sqrt{13}$
41. $(3 \pm 3\sqrt{17})/2$
43. $(-3 \pm 3\sqrt{73})/4$
45. $(11 \pm \sqrt{305})/4$

Exercise Set 5.2

1. If a leg and an acute $\angle$ of one right $\triangle$ are $\simeq$ to the corr. parts of another right $\triangle$, then the two right $\triangle$s are $\simeq$. Use ASA = ASA or SAA = SAA.
3. If the two legs of one right $\triangle$ are $\simeq$ to the corr. parts of another right $\triangle$, then the two right $\triangle$s are $\simeq$. Use SAS = SAS.
5. Use HL = HL.
7. Use HA = HA.
9. Use LL = LL.
11. Show that $AF = BG$ so that $\triangle ADF \simeq \triangle BEG$ by LA = LA. Thus, $DF = EG$. Hence, $\triangle DFC \simeq \triangle EGC$ by LL = LL. Therefore, $DC = EC$ and the result follows by Theorem 2.4.
13. Show that if $\triangle ABC \simeq \triangle DEF$, then $\triangle DEF \simeq \triangle ABC$.

Exercise Set 5.3

1. $27/2$
3. $15/4$
5. $8\sqrt{39}/5$
7. $2\sqrt{3}$
9. $\triangle AFB, \triangle CED, \triangle ABC, \triangle CDA, \triangle BFC, \triangle DEA$
11. Use Theorem 3.5 and Theorem 5.2.

Exercise Set 5.4

1. 13
3. $144/13$
5. $\sqrt{5}$
7. 10
9. $5\sqrt{5}$
11. 3
13. 5
15. $\sqrt{253}/5$
17. $2\sqrt{3}$
19. $4\sqrt{3}$
21. $15°$
23. $30°$
25. $4\sqrt{3}$
27. $80/9$
29. In an isos. right $\triangle$ the two legs are =. Thus, by the Pyth. Theorem, the square of the length of the hypotenuse equals twice the square of the length of one leg. The result follows.
31. Use Theorem 5.6 and the Pyth. Theorem.
33. $800 - 100\sqrt{34} \approx 217$ ft
35. $2 + 10\sqrt{58} \approx 78$ ft

Exercise Set 5.5

1. $\sqrt{3}/2$
3. $1/\sqrt{2}$
5. $1/2$
7. $\sqrt{3}$
9. $1/\sqrt{3}$
11. $8/17$
13. $8/17$
15. $8/15$
17. $2/\sqrt{13}$
19. $2/\sqrt{13}$
21. $2/3$
23. 0.9563
25. 0.9816

27. 5.145
29. 0.8387
31. 0.9976
33. 14.30
35. 0.7547, 0.7547
37. 0.9998, 0.9998
39. Approx. 57 ft
41. Approx. 15432 ft
43. Approx. 11 ft
45. Approx. 1750 miles
47. Approx. 193 ft, 187 ft
49. $\dfrac{\sin m\angle B}{\tan m\angle B} = \dfrac{OPP/HYP}{OPP/ADJ} = \dfrac{OPP}{HYP} \cdot \dfrac{ADJ}{OPP}$

$\qquad = \dfrac{ADJ}{HYP} = \cos m\angle B$

Chapter 5 Review Exercises

1. Sometimes
3. Sometimes
5. Always
7. Never
9. Never
11. Similar
13. True
15. True
17. True
19. True
21. 6
23. $(\sqrt{5}/2)BC$
25. $3\sqrt{2}/2$
27. $6\sqrt{2}$
29. 10/3
31. 23/9
33. 6
35. 17
37. $a\sqrt{41}$
39. $\sqrt{39}$
41. $\sqrt{6}/6$
43. $n\sqrt{33}$
45. $\{5a, 12a, 13a\}$, $\{7a, 24a, 25a\}$, $\{8a, 15a, 17a\}$, $\{9a, 40a, 41a\}$, etc., for any positive integer a
47. 1/2
49. 37°
51. 9/25
53. $12\sqrt{5}$
55. $5\sqrt{3}/6$ parsecs
57. 1.5 centimeters

59. Show that $\angle RNQ \simeq \angle SNQ$. Then use HA = HA.
61. Use the Corollary to Theorem 5.6.
63. Show that $\triangle ADF \simeq \triangle BEG$ by SAS = SAS and thus, $\angle A \simeq \angle B$. The result follows from Theorem 2.5.

Exercise Set 6.1

1. Tangent segment
3. Tangent ray
5. Radius
7. Internally tangent
9. Concentric
11. Nonintersecting
13. $8 + 16\sqrt{3}/3$.
15. Show that $\triangle PAC \simeq \triangle PBC$ by HL = HL.
17. Show that $\overline{PB} \perp \overline{BC}$ and $\overline{QD} \perp \overline{BC}$.
19. If d is the distance from the center of the circle to the line, then $d < r$ (radius). Use segments of length $\sqrt{r^2 - d^2}$ on the line to show that there are two points on the line that are at a distance r from the center of the circle.
21. $200\sqrt{41}$ miles
23. $300\sqrt{5}$ miles
27. Umbra

Exercise Set 6.2

1. =
3. <
5. $2 + 4\sqrt{2} + 2\sqrt{3}$
9. Show $\triangle PAQ$ is isosceles. Thus, $\angle APQ \simeq \angle AQP$ and hence, $\angle BAP \simeq \angle CAQ$. Also $\triangle ABP$ and $\triangle ACQ$ are isosceles. The result follows.
11. Show that $\triangle ADB \simeq \triangle CDB$ using LL = LL.
13. Use the perpendicular bisectors of the line segments joining the points. These will intersect at the center of the required circle.
15. Use SSS = SSS and use the def. of $\perp$.

Exercise Set 6.3

1. $\overset{\frown}{EF}, \overset{\frown}{FG}, \overset{\frown}{GH}, \overset{\frown}{HE}, \overset{\frown}{EHG}, \overset{\frown}{FGH}$
3. One ray of $\angle S$ intersects $\odot Q$ only at the vertex. Both rays of $\angle T$ intersect $\odot Q$ only at the vertex.
5. 266°

7. 180°

9. 333°

11. 120°

13. 65°

15. 69°

17. 90°

19. 25°

21. 70°

23. 45°

25. Similar to Case 1.

27. Use Theorem 6.10.

29. Use Theorem 6.2 and Theorem 6.10.

31. The statements are identical to that of Theorem 6.13 except that in the first diagram one of the rays is a tangent, and in the second diagram, both rays are tangents. Proofs are similar to the proof of Theorem 6.12.

Exercise Set 6.4

1. 9

3. 4, 10

5. $8\sqrt{3}$

9. Use AA ~ AA.

11. Show that $\triangle ABP \simeq \triangle ACP$ by LL = LL. Thus, $\angle BAD \simeq \angle CAD$.

13. Given chords $\overline{AB}$ and $\overline{CD}$ such that $CD > AB$, construct a circle P with chords $\overline{EF}$ and $\overline{FG}$ such that $EF = AB$ and $FG = CD$ and such that E is not on the same side of $\overleftrightarrow{GF}$ as P. Now $m\overset{\frown}{FE} + m\overset{\frown}{EG} = m\overset{\frown}{FG}$ and $m\overset{\frown}{EG} > 0$, so that $m\overset{\frown}{FG} > m\overset{\frown}{FE}$. Therefore $m\overset{\frown}{CD} > m\overset{\frown}{AB}$.

15. Show that $\triangle ADC \sim \triangle ABE$ by AA ~ AA.

17. 4800 miles

Exercise Set 6.5

11. Show that $\overleftrightarrow{PA} \perp \overleftrightarrow{AB}$ and $\overleftrightarrow{QB} \perp \overleftrightarrow{AB}$. Then use Theorem 6.1.

Exercise Set 6.6

1. The ray that bisects $\angle A$

3. Two parallel lines 4 inches apart

5. The center of the circle containing the three points

7. A line parallel to m and n, half an inch from each of them

9. Two points, one point, or $\emptyset$

11. A diameter of the circle (excluding the endpoints)

13. Four concentric circles, of radius 1, 2, 4, and 5

15. The union of (a) a figure in the exterior of the pentagon that looks like a pentagon except that instead of vertices there are arcs, and (b) a pentagon similar to the original pentagon, (or a point or $\emptyset$) in the interior of the pentagon.

Chapter 6 Review Exercises

1. Sometimes

3. Never

5. Sometimes

7. Never

9. Always

11. True

13. Is inscribed in

15. True

17. Twice

19. True

21. 10

23. 3 (or 5 if you don't believe $DF < FG$)

25. 2°30'

27. Theorem 6.7

29. Theorem 6.2

31. Arc Add. Theorem

33. Let $\overline{PR} \cap \overline{LS} = \{U\}$. Then $PU > PS$ by Theorem 3.3 and $PR > PU$ by def. of between. Therefore, $PR > PS$.

35. The union of (a) a figure in the exterior of the rectangle that looks like a rectangle except that instead of vertices there are arcs, and (b) a line segment one inch long in the interior of the rectangle.

37. Two perpendicular lines

Exercise Set 7.1

1. Yes

3. No

5. Yes

7. 6

9. 18

11. 0 (no region)

13. 30 in^2

15. $3\sqrt{2}$ in, 6 in
17. $\sqrt{10}$ cm, $2\sqrt{5}$ cm
19. 1/9
21. 72
23. Divide the polygons into triangles and use Postulate 7.2.

Exercise Set 7.2

1. 10
3. 75
5. 22
7. 350 in^2
9. 7.5
11. 8 cans (582 ft^2)
13. 36.8125 in^2
15. $288\sqrt{3}$ ft^2
17. $7\sqrt{30}$
19. $6\sqrt{10}$
25. Use Theorem 7.3.
27. Divide the polygon into triangles formed by the central angles and the sides and use Theorem 7.3.
29. Use Theorem 7.3 and Postulate 7.3.
31. Use Theorem 7.3.

Exercise Set 7.3

1. 4π
3. $8\pi/3$
5. $4\pi/3 - \sqrt{3}$
7. 4π
9. $8\pi/3 + 4$
11. $4\pi/3 + 2\sqrt{3}$
13. $144 - 36\pi$
15. $8\pi + 16\sqrt{3}$
17. 75π
19. $10\pi + 4\sqrt{3}$, 10π
21. Use Theorem 7.6.
23. Use diagrams.
25. 5

Chapter 7 Review Exercises

1. Sometimes
3. Never
5. Never

7. a^2/b^2
9. True
11. $4\sqrt{3}$, $2\sqrt{3}$
13. $16\pi/3 - 4\sqrt{3}$
15. $\sqrt{3} - 1$
17. $64\pi - 96\sqrt{3}$
19. Use Theorem 7.2.
21. Area $= a(b_1 + b_2)/2$
23. $26\pi + 48$
25. $100\pi - 96$
27. $(59\pi - 16\sqrt{3})/3$

Exercise Set 8.1

1. 5
3. 25
5. $\sqrt{89}$
7. $16\sqrt{2}$
9. Yes, no, yes, no
11. Yes, yes, yes, yes
13. $(-3, -8)$
15. $(0, 7)$
17. Consider the points $P(a, b)$, $Q(c, d)$, and $R(c, b)$ and show that $(PR)^2 + (QR)^2 = (PQ)^2$.

Exercise Set 8.2

1. 1
3. 2
5. 7/4
7. (9/2, 15/2), (3/2, 1), (17/2, 1), (11/2, −2), (2, 7/2), (5, 2)
9. (9, 11)
11. −2, 3/2
13. 3, −17
15. 1/7, 6/7
17. $x + 3y - 24 = 0$
19. $2x - 5y + 24 = 0$
21. $x = -7$

Exercises 23, 25, 27: draw line segments from:

23. (0, −8) to (3, 4)
25. (−3, 0) to (3, 8)
27. (4, −6) to (4, −2)
29. $18x + 13y - 68 = 0$, $21x - 4y - 41 = 0$, $3x - 17y + 27 = 0$, (7/3, 2)

Exercise Set 8.3

1. $2x - 3y + 18 = 0, [-6, 3]$;
$9x + 5y + 44 = 0, [-6, -1]$;
$15x - 4y - 13 = 0, [-1, 3]$
3. $5x - 3y + 30 = 0, [-6, 0]$;
$5x + 6y - 60 = 0, [0, 6]$;
$y = 5, [0, 6], x - y + 5 = 0, [-3, 0]$;
$2x - 3y + 12 = 0, [-6, -3]$
5. $5x + 2y + 11 = 0, [-3, -1]$;
$2x - 5y + 16 = 0, [-3, 2]$;
$5x + 2y - 18 = 0, [2, 4]$;
or $5x + 2y + 11 = 0, [-1, 1]$;
$2x - 5y - 42 = 0, [1, 6]$;
$x + 2y - 18 = 0, [4, 6]$

9. Consider the triangle with vertices $A(a, 0), B(b, d)$, and $C(0, 0)$. The midpoint of $\overline{AB}$ is $M((a+b)/2, d/2)$ and the midpoint of $\overline{BC}$ is $N(b/2, d/2)$. Notice that $\overline{AB} \| \overline{BC}$, since both are on lines with slope zero. Using Theorem 8.1, we find that $AC = a$ and $MN = (1/2)a$, yielding the desired result.

11. Consider the triangle with vertices $A(a, 0), B(b, d)$, and $C(0, 0)$. The midpoint of $\overline{AB}$ is $M((a + b)/2, d/2)$, the midpoint of $\overline{BC}$ is $N(b/2, d/2)$, and the midpoint of $\overline{AC}$ is $P(a/2, 0)$. The slope of $\overleftrightarrow{CM}$ is $d/(a + b)$, the slope of $\overleftrightarrow{AN}$ is $d/(b - 2a)$. The equation of $\overleftrightarrow{CM}$ is $y = dx/(a+b)$ and the equation of $\overleftrightarrow{AN}$ is $y = d/(x - a)(b - 2a)$. Solving these two equations in two unknowns yields $x = (a + b)/3$ and $y = d/3$, so that the point of intersection of $\overline{CM}$ and $\overline{AN}$ is $((a + b)/3, d/3)$, which is two-thirds of the distance from A to N and from C to M. Similar reasoning yields the same intersection point on $\overline{BP}$.

Exercise Set 8.4

1. $x^2 + y^2 = 9$
3. $(x + 4)^2 + (y + 7)^2 = 16$
5. $x^2 + y^2 - 8x + 16y - 89 = 0$
7. $x^2 + y^2 + 6x - 12y - 44 = 0$
9. $x^2 + y^2 - 8x + 12y + 36 = 0$
11. $(2, 5), 3$
13. $(-7, 25/2), \sqrt{593}/2$
15. $x^2 + y^2 + 6x - 8y + 16 = 0$,
$x^2 + y^2 + 6x - 8y + 9 = 0$,
$x^2 + y^2 + 6x - 8y = 0$,
$x^2 + y^2 + 6x - 8y - 11 = 0$

Exercise Set 8.5

1. Yes
3. Yes
5. Yes
7. No
9. No
11. Yes
13. Yes
15. Yes
17. Yes
19. No
21. Yes
23. No
25. No
27. Yes
29. No
31. Yes
33. Both yield the same result. Yes.
35. The results are different.
37. The results are different.
39. Fourth: translate $\mathbb{R}$ 1 unit up, then reflect about the line $y = -x$, and then rotate 45° clockwise. Fifth: rotate $\mathbb{R}$ 45° clockwise, then reflect about the line $y = -x$, and then translate 1 unit up. Sixth: reflect $\mathbb{R}$ about the line $y = -x$, then translate 1 unit up, and then rotate 45° clockwise. The third order yields the same set of points as the fifth order. The other results are different.

41. *Given:* $\mathcal{T} : \mathbb{R} \leftrightarrow \mathbb{S}$ is an isometry;
A, B, C in $\mathbb{R}$; A', B', C' in $\mathbb{S}$;
$A \leftrightarrow A', B \leftrightarrow B', C \leftrightarrow C'$;
B is between A and C
Prove: B' is between A' and C'
Proof: Since B is between A and C, $AB + BC = AC$. Since $\mathcal{T}$ is an isometry, it preserves distances, so that $AB = A'B', BC = B'C'$, and $AC = A'C'$. Thus, $A'B' + B'C' = A'C'$ by substitution. Therefore, B' is between A' and C'.

43. *Given:* $\mathcal{T} : \mathbb{R} \leftrightarrow \mathbb{S}$ is an isometry;
A, B, C in $\mathbb{R}$; A', B', C' in $\mathbb{S}$;
$A \leftrightarrow A', B \leftrightarrow B', C \leftrightarrow C'$; $\triangle ABC$,
$\triangle A'B'C'$
Prove: $\triangle ABC \simeq \triangle A'B'C'$
Proof: Since $\mathcal{T}$ is an isometry, it preserves distances, so that $AB = A'B', BC = B'C'$, and $AC = A'C'$. Therefore, $\triangle ABC \simeq \triangle A'B'C'$ by SSS = SSS.

45. *Given:* $\mathcal{T} : \mathbb{R} \leftrightarrow \mathbb{S}$ is a reduction;
     $A(x_1, y_1), B(x_2, y_2), C(x_3, y_3)$ in $\mathbb{R}$;
     $A'(rx_1, ry_1), B'(rx_2, ry_2), C'(rx_3, ry_3)$
     in $\mathbb{S}$; $A \leftrightarrow A', B \leftrightarrow B', C \leftrightarrow C'$;
     $\triangle ABC, \triangle A'B'C'$
  Prove: (a) $\triangle ABC \sim \triangle A'B'C'$
      (b) $\triangle ABC \neq \triangle A'B'C'$
  Proof: (a) Show that $AB/A'B' = BC/B'C' = $
      $AC/A'C' = r$. Then use SSS $\sim$ SSS.
      (b) Show that $AB \neq A'B'$.

47. $\mathbb{S} = \mathbb{R}$, so that $x^2 + y^2 = 1$. Thus, every point $\mathbb{R}$ with coordinates (x, y) corresponds to the point in $\mathbb{S}$ with coordinates (x, y). Therefore, $\mathcal{T}$ is a transformation. Since coordinates are preserved, clearly distances are preserved. Therefore, $\mathcal{T}$ is an isometry.

49. $\mathbb{S}$ is the exterior of the unit circle. (Notice that $x^2 + y^2 < 1$, so that x divided by this quantity is greater than x and y divided by this quantity is greater than y). $\mathcal{T}$ is a transformation because the one-to-one correspondence determined by the coordinates of the points yields a one-to-one correspondence between the points. We can show that $\mathcal{T}$ is not an isometry by showing that distance is not always preserved. For example, the distance between $(0, 0.5)$ and $(0, 0.1)$ is 0.4, but the distance between the corresponding points $(0,2)$ and $(0,10)$ is 8.

Chapter 8 Review Exercises

  1. Never
  3. Sometimes
  5. Sometimes
  7. Undefined
  9. Are not
  11. $\sqrt{305}$
  13. $(-5\sqrt{2}, 5\sqrt{2}), (5\sqrt{2}, -5\sqrt{2})$
  15. $3x - 2y + 4 = 0, [-6, 2]$
  17. $3/5, -11/5$
  19. Show that the opposite sides of the quadrilateral have the same slope
  21. Consider the right triangle with vertices at $A(a, 0)$, $B(0, b)$, and $O(0, 0)$, and the midpoint $M(a/2, b/2)$ of the hypotenuse. Show that $OM = (1/2)AB$.
  23. Yes
  25. No

Exercise Set 9.1

  3. One
  5. No

  7. Yes
  9. T intersects neither R nor S.
  11. One, infinitely many.
  13. Consider a right triangle with sides 3, 4, and 5.

Exercise Set 9.2

  1. 4, 4, 6
  3. 8, 6, 12
  5. 20, 12, 30
  7. 3, 3
  9. 4, 4
  11. 0
  13. 100
  19. 7, 10

Exercise Set 9.3

  1. $9\sqrt{3}$ square meters, $18\sqrt{3}$ square meters
  3. 460 square inches
  5. $5\pi\sqrt{89}$ square centimeters
  7. 140π square inches
  9. $64,000,000\pi$ square miles
  11. $108 + 18\sqrt{3}$ square inches

Exercise Set 9.4

  1. 420 cubic meters
  3. 27 to 1
  5. 1152π cubic meters
  7. $40\sqrt{3}$ cubic centimeters
  9. $\pi r^3/3$
  11. $256,000,000,000\pi/3$ cubic miles
  13. 6π cubic inches
  15. $212\pi/3$ cubic feet

Chapter 9 Review Exercises

  1. Sometimes
  3. Never
  5. Sometimes
  7. Sometimes
  9. Sometimes
  11. True
  13. Skew
  15. Pentagonal
  17. True
  19. Must
  21. One
  23. 64

25. $25\sqrt{3}$ square centimeters
27. Cone (120π versus $1000/3$ cubic inches)
29. 522π square feet
31. $127/27$ meters
33. 435 cubic yards
35. $1,072,476,000,000,000,000,000,000\pi$ cubic miles

Exercise Set 10.1

1. Infinitely many
3. One
5. One
7. Less than $180°$
9. The measure of an exterior angle of a triangle is greater than the sum of the measures of the opposite interior angles.

Exercise Set 10.2

1. None
3. Three

5. Depending upon the position of the point, there may be one or infinitely many.
7. Between $180°$ and $540°$
9. The measure of the exterior angle of a triangle is less than the sum of the measures of the opposite interior angles.

Chapter 10 Review Exercises

1. *S*
3. *E, P*
5. *S*
7. *E*
9. *S*
11. Hyperbolic: less than $180°$; elliptic: between $180°$ and $540°$
15. Consider numbers of the form $1/r$, in particular when r is near zero.

INDEX

List of Abbreviations

add.	addition	meas.	measure
adj.	adjacent	mult.	multiplication
alt.	altitude; alternate	opp.	opposite
approx.	approximately	post.	postulate
cm	centimeter(s)	pt., pts.	point, points
comp.	complementary	quad.	quadrilateral
corr.	corresponding	seg.	segment
cos	cosine		ne
def.	definition		aight
diag.	diagonal		tement
ft	foot (feet)		pplementary
geom.	geometric		gent
hyp.	hypotenuse		nsversal
in.	inch(es)		tical